# Cutting-Edge Mobile Applications for Higher Education

Mohammed Amin Almaiah
https://orcid.org/0009-0008-9785-485X
*The University of Jordan, Amman, Jordan*

Said Salloum
*School of Science, Engineering, and Environment, University of Salford, UK*

IGI Global
Scientific Publishing
Publishing Tomorrow's Research Today

Published in the United States of America by
IGI Global Scientific Publishing
701 East Chocolate Avenue
Hershey, PA, 17033, USA
Tel: 717-533-8845 | Fax: 717-533-7115
Website: https://www.igi-global.com E-mail: cust@igi-global.com



Library of Congress Cataloging-in-Publication Data

Names: Almaiah, Mohammed Amin, 1986- editor | Salloum, Said Abdelrahim, 1978- editor
Title: Cutting-edge mobile applications for higher education / edited by: Mohammed Amin Almaiah and Said Salloum.
Description: Hershey, PA : IGI Global Scientific Publishing, [2026] | Includes bibliographical references and index. | Summary: "The book aims to continue and promote the discussion about smart technologies supported learning environments in higher education and therefore invites original papers focusing on the latest developments in smart and mobile educational technologies"-- Provided by publisher.
Identifiers: LCCN 2025009708 (print) | LCCN 2025009709 (ebook) | ISBN 9798369382875 hardcover | ISBN 9798369382882 paperback | ISBN 9798369382899 ebook
Subjects: LCSH: Educational technology | High technology and education | Mobile apps
Classification: LCC LB1028.3 .C88 2025 (print) | LCC LB1028.3 (ebook) | DDC 371.33--dc23/eng/20250813
LC record available at https://lccn.loc.gov/2025009708
LC ebook record available at https://lccn.loc.gov/2025009709

British Cataloguing in Publication Data
A Cataloguing in Publication record for this book is available from the British Library.

# Table of Contents

# Detailed Table of Contents



The chapter explore the impact of mobile and smart technologies on higher education systems, assessing AI-powered environments worldwide, with special reference to India. It also evaluates the role of these applications in enhancing communication, collaboration, and peer learning in higher education while projecting the growth of global networks of education. It even highlights the benefits, challenges, and future trends of AI, IoT, and adaptive learning in academics. In addition, it reviews strategies related to policies, case studies, and implementation which influence the adoption of mobile learning technologies in educational institutions worldwide.



The integration of Artificial Intelligence (AI) and Machine Learning (ML) in educational applications is transforming higher education by enhancing personalized learning, intelligent tutoring, and predictive analytics. This chapter explores AI-driven functionalities, including adaptive learning, NLP, and automated assessments, while addressing challenges such as data privacy, algorithmic bias, and accessibility. Through case studies, it highlights AI's transformative potential in shaping future education, offering insights for educators, researchers, and developers interested in AI-driven learning innovations..

In the technological era, smart and mobile applications are playing a crucial role in enhancing the learning experience of students to the next level. To understand its impact on higher education, this chapter explores the application of Augmented Reality (AR) and Virtual Reality (VR) to build immersive learning environments, as well as the integration of Artificial Intelligence (AI) to customize educational content to each student's needs. It also addresses the challenges associated with mobile learning, such as the digital divide, security concerns, and the need for effective pedagogy. Furthermore, the chapter discusses future trends, including the rise of micro learning, gamification, and the impact of 5G technology. By examining case studies and success stories, the chapter provides a comprehensive overview of how smart and mobile applications are reshaping higher education, ultimately seeking to establish a more welcoming learning environment.

# Preface

## INTRODUCTION

Welcome to a book "Cutting-Edge Mobile Applications for Higher Education", a comprehensive compilation edited by Mohammed Amin Almaiah and Said Salloum. In this volume, our aim is to provide a systematic exploration of essential mobile technologies in higher education concepts, methodologies, and technologies.

The rapid evolution of mobile technologies has profoundly transformed the landscape of higher education. Smartphones, tablets, and wearable devices have become integral to academic life, enabling anytime, anywhere access to learning resources, institutional services, and collaborative platforms. Beyond basic communication tools, modern mobile applications now incorporate advanced capabilities such as artificial intelligence, augmented and virtual reality, learning analytics, cloud computing, and secure authentication mechanisms. These innovations are redefining how teaching, learning, assessment, and academic administration are designed and delivered.

In higher education institutions, mobile applications play a critical role in supporting student engagement, personalized learning, and institutional efficiency. Mobile learning management systems, virtual laboratories, adaptive learning apps, and smart campus solutions facilitate interactive learning experiences and foster data-driven decision-making. At the same time, challenges related to usability, accessibility, privacy, security, scalability, and ethical use of data remain significant concerns for educators, developers, and policymakers.

The growing demand for resilient, secure, and intelligent mobile solutions particularly in response to digital transformation initiatives and post-pandemic educational models and highlights the need for a comprehensive academic reference that bridges theory, design, and real-world implementation. This book, Cutting-Edge Mobile Applications for Higher Education, addresses this need by presenting state-of-the-art research, practical frameworks, and case studies that explore the development, deployment, and evaluation of advanced mobile applications within higher education ecosystems.

We extend our gratitude to all the contributors who have shared their expertise and insights, helping to enrich the discourse surrounding the adoption of mobile learning in higher education. It is our sincere hope that this book will foster further collaboration and innovation in the ongoing quest to enhance our digital learning.

## ORGANIZATION OF THE BOOK

In "*The Transformative Ramifications of Smart and Mobile Technologies on Higher Education*" by *Uma, Monica Verma and Divya Mathur*. The chapter explores the impact of mobile and smart technologies on higher education systems, assessing AI-powered environments worldwide, with special reference to India. It also evaluates the role of these applications in enhancing communication, collaboration, and peer learning in higher education while projecting the growth of global networks of education. It even highlights the benefits, challenges, and future trends of AI, IoT, and adaptive learning in academics. In addition, it reviews strategies related to policies, case studies, and implementation that influence the adoption of mobile learning technologies in educational institutions worldwide.

In "*Artificial Intelligence and Machine Learning In Educational Apps*" by *Alwin T. George, Varuna Gupta, Vandana Mehndiratta*. This chapter explores AI-driven functionalities, including adaptive learning, NLP, and automated assessments, while addressing challenges such as data privacy, algorithmic bias, and accessibility. Through case studies, it highlights AI's transformative potential in shaping future education, offering insights for educators, researchers, and developers interested in AI-driven learning innovations.

In "*Enhancing Higher Education with a Mobile Application: Leveraging Advanced Technologies for Personalized Learning, Career Guidance, and Accessibility*" by *C.V. Suresh Babu, Anjana Priya S S, Benzen A, Vignesh A*. This study aims to develop a mobile application that enhances the higher education experience by addressing key challenges such as accessibility, engagement, and personalization. Utilizing a user-centered design approach and agile development methodology, the research involved iterative cycles of design, testing, and feedback from stakeholders, including students and educators. The findings indicate that the application significantly improves student engagement through interactive features and personalized learning pathways, while also ensuring data security and privacy. The implications of this research highlight the potential of mobile technology to democratize education, foster lifelong learning, and prepare students for a rapidly evolving job market. Ultimately, the mobile application serves as a transformative tool in higher education, promoting a more inclusive and effective learning environment.

In "*Game-based learning and gamification*" by *Tanya Shakya and Nikita Joshi*. This chapter invites readers to explore the exciting potential of these innovative methodologies to enhance educational outcomes, particularly in fields that demand practical application and critical thinking, such as cybersecurity and technology. Game-based learning transforms traditional education by turning the learning process itself into an engaging game. In this approach, carefully crafted games are designed to deliver educational content, tapping into the innate motivation that game mechanics inspire. Students are encouraged to dive into interactive worlds where they can explore, experiment, and learn through hands-on experiences and trial-and-error scenarios. Game-based learning is a fun and effective way to keep students engaged while helping them remember what they learn.

In "*Innovative Pedagogies in Online and Blended Learning Environments: Transforming Education in the Digital Age*" by *Seema Yadav*. This chapter explores the transformative impact of online and blended learning on educational practices, emphasizing the theoretical frameworks, advantages, methodologies, assessment strategies, and future trends shaping this dynamic landscape. It highlights the role of constructivist and cognitive theories in designing learner-centered environments that promote engagement and collaboration. It examines the pedagogical innovations, technological advancements, and instructional strategies that are redefining the educational experience. The chapter provides a comprehensive analysis of the benefits, challenges, and future directions of online and blended learning, drawing on recent research and case studies to highlight effective practices and emerging trends.

In "*Mobile Learning and Digital tools Effectiveness in Higher Education System in India*" by *Vijay Ganpat Dhamore*. This study will focus on higher education and the Digital tools or Mobile Apps used to teaching and learning monitoring systems in India for the Higher Education systems Purposive Sampling Method used for conducting the study. Primary Data collected through a Standard Questionnaire form and with the help of google form. Students from the Higher Education Students were respondents. Data analysis tools like SPSS and SEM used to analyze the data. This study finally concludes for scope of LMS and Mobile Application and its effectiveness in Current Higher Education Systems in India.

In "*Next Gen Education: Elevating Higher Learning With Smart and Mobile Applications*" by *Surenthran David P, Daniel Thangadurai Hepzibah Vinsyah Jeyaseeli, Joel Jebadurai Devapitchai, M Nanda kumar, M M Gayathri, S Vijayalakshmi*. This chapter explores the application of Augmented Reality (AR) and Virtual Reality (VR) to build immersive learning environments, as well as the integration of Artificial Intelligence (AI) to customize educational content to each student's needs. It also addresses the challenges associated with mobile learning, such as the digital divide, security concerns, and the need for effective pedagogy. Furthermore, the chapter discusses future trends, including the rise of micro learning, gamification, and the impact of 5G technology. By examining case studies and success stories, the chapter provides a comprehensive overview of how smart and mobile applications are reshaping higher education, ultimately seeking to establish a more welcoming learning environment.

## IN CONCLUSION

This book "Cutting-Edge Mobile Applications for Higher Education" has provided a comprehensive exploration of cutting-edge mobile applications for higher education by bringing together theoretical foundations, technological innovations, design methodologies, and real-world implementations. The chapters collectively highlight how intelligent, secure, and user-centered mobile solutions can support personalized learning, foster student engagement, enable smart campus ecosystems, and improve decision-making through data-driven insights. At the same time, the book points out urgent issues related to security, privacy, accessibility, and ethical use, emphasizing the need for responsible and sustainable mobile application development.

Looking ahead, the future of mobile applications in higher education will be shaped by increasing interoperability, adaptive learning systems, and greater alignment with institutional strategies and educational policies. To fully realize the potential of mobile technologies and address emerging risks and constraints, researchers, educators, developers, and policymakers must engage in continuous collaboration. By serving as a multidisciplinary reference, Cutting-Edge Mobile Applications for Higher Education aims to guide future research, inform best practices, and inspire innovative solutions that support inclusive, secure, and effective higher education environments.

As we conclude this preface, we extend our sincere appreciation to all the authors who have contributed their expertise and insights to this volume. It is our hope that "Cutting-Edge Mobile Applications for Higher Education" will serve as a valuable resource for years to come, inspiring further exploration and innovation in the ongoing mission to enhance our digital learning.

**Mohammed Amin Almaiah**

*The University of Jordan, Amman, Jordan*

**Said Salloum**

*School of Science, Engineering, and Environment, University of Salford, UK*

Chapter 1

# The Transformative Ramifications of Smart and Mobile Technologies on Higher Education

**Uma**
https://orcid.org/0000-0001-5287-6803
*Darshan University, India*

**Monica Verma**
https://orcid.org/0009-0004-6292-8979
*Marwadi University, India*

**Divya Mathur**
https://orcid.org/0000-0001-6083-8487
*Marwadi University, India*

## ABSTRACT

*The chapter explore the impact of mobile and smart technologies on higher education systems, assessing AI-powered environments worldwide, with special reference to India. It also evaluates the role of these applications*

DOI: 10.4018/979-8-3693-8287-5.ch001

*in enhancing communication, collaboration, and peer learning in higher education while projecting the growth of global networks of education. It even highlights the benefits, challenges, and future trends of AI, IoT, and adaptive learning in academics. In addition, it reviews strategies related to policies, case studies, and implementation which influence the adoption of mobile learning technologies in educational institutions worldwide.*

## 1. INTRODUCTION

The exponential growth in mobile and smart technologies have significantly transformed higher education by introduction to new pedagogical models and interactive learning environment which can be globally accessed. There are major changes due to AI, IoT, cloud computing, and adaptive learning applications which has reshaped delivery of education. Worldwide universities have accepted AI-driven platforms and mobile learning management systems (LMS) to facilitate adaptive learning environments (Bacca et al., 2014; Spector, 2019). Moreover, mobile learning (m-learning) applications have bridged geographic as well as socio-economic barriers, providing real-time access to education. (Traxler, 2018; Kumar & Sharma, 2020).

The present paper explores the evolution and history, adoption and effect of AI-driven and smart mobile learning applications in higher education by focusing on their role in communication, collaboration, peer learning and global partnerships.

The chapter has highlighted the developments smart and mobile learning technologies from early e-learning to AI-based adaptive learning systems at present. It analyses AI-driven mobile applications, trends and role in their global acceptances and academic partnership, integrating Augmented Reality (AR), Virtual Reality (VR), Internet of Things (IoT), and smart campuses. Further, it discusses various challenges involved within like privacy of data, algorithmic biases, ethical issues and the digital divide, while highlighting developing inventions such as 5G-powered block-chain technology and learning. The chapter is structured into *seven categories* for a comprehensive analysis. Finally, it discusses regulatory policy frameworks, their implementation, challenges involved and the best practices in integrating mobile learning technologies into mainstream education.

## 2. EVOLUTION OF SMART AND MOBILE TECHNOLOGIES IN HIGHER EDUCATION

### 2.1 Historical Development

The initial phase of evolution of educational technology can be witnessed from chalk and duster to AI-driven smart learning platforms. This switch from traditional LMS to AI-driven Intelligent Tutoring Systems (ITS) has improved engagement and performance of the students. Moreover, with the emergence of 5G connectivity, cloud-computing, and augmented reality (AR) enhanced real-time access to high-quality educational content has become possible.

This advancement has significantly transformed methods of how students learn and also how educators instruct. This change from conventional Learning Management Systems (LMS) to AI-powered Intelligent Tutoring Systems (ITS) has improved students' engagement and performance. (Holmes et al., 2021). During 1980s with the introduction to computers, doors were opened to computer-assisted learning programs & there was a revolutionary effect of the internet on the delivery of education which was being considered as a major transformation (U.S. Department of Education, 2023). Radio, TV, and film projectors alike technologies played a crucial role in growing the access to education that was beyond classrooms (O'Shea, 2023).

Reshaping of the landscapes of higher education with the fast growth of smart and mobile technologies since it introduced new pedagogical models, which create an interactive and participative learning environments accessible to global community. (Al-Hunaiyyan, Alhajri, & Bimba, 2020). These technologies have made reach of higher education possible not only personalized but beyond traditional classroom settings which include distance learning, virtual learning, and real time feedback mechanisms. (Spector, 2019). Competency based and adaptive learning have been emerged through the smart learning environments—driven by AI-based platforms, cloud computing, and mobile-based learning management systems (LMS. (Bacca et al., 2014). Further, mobile learning applications (m-learning) have revolutionized acquisition of knowledge, breaking the traditional geographical and socio-economic barriers. (Traxler, 2018).

## 2.2 The Role of AI in Personalized Learning

Smart Learning Environments (SLEs) which are AI-based, support personalized learning and data-driven assessments by generating machine learned algorithms and providing analytics to track student's performance, identify gaps, and optimize learning pathways. Experiences with technologies such as AI-driven tutors, chatbots, immersive Virtual Reality (VR) and Augmented Reality (AR) provide a real-time support and individualized learning by enhancing engagement and comprehension (Holmes et al., 2021; Billinghurst & Duenser, 2012).

Cloud computing became faster with the advent of 5G technology and students now can access to the high quality content, participate in online discussions through various online forums, collaborate on their research projects, anytime and anywhere (Kumar & Sharma, 2020). AI-driven mobile platforms provide with the real-time feedback making it more personalized. Also, tailored educational pathways are suggested, to improve the retention and performance rates in academics (Siemens & Long, 2011; Laak & Aru, 2024).

AI-powered tools and intelligent tutoring systems such as Carnegie Learning and AI-enabled chatbots, enhance individualized learning, whereas, Grammarly and Turnitin like application provide prompt feedback on originality, grammar, coherence. (VanLehn, 2011). However, there are existing challenges of privacy, ethical issue, and digital divide which must be addressed so that equal access to AI-driven education can be ensured. (Grant-Kels, 2017; Al-Kahtani et al., 2021). Researchers highlight the need for ethical and responsible AI integration strategies to maintain student privacy and ensure responsible usage of AI (Pedro et al., 2019). Case studies of leading universities highlight that AI-driven personalized models of learning are improving academic outcomes. (Siemens & Long, 2011)

# 3. AI-POWERED MOBILE APPLICATIONS AND THEIR ROLE IN EDUCATION

## 3.1 Personalized Learning and Adaptive Systems

Adaptive learning models and intelligent tutoring systems have brought major changes education by facilitating with real-time assessments which lead to customized learning pathways. (Zawacki-Richter et al., 2023). The data-driven insights in the form of algorithms based on machine learning models provide analysis of students' behaviour. As a result, tailoring content to their individual needs can be undertaken. Further, improving retention and engagement among the students. (Laak & Aru, 2024). To experience an immersive and experienced learning AI and AR are integrated on cloud based adaptive learning platforms. (Marienko et al., 2020).

Mobile learning based on AI models enahance engagement, support hybrid models of education, and provides real-time assessments. Carnegie Learning based on intelligent tutoring systems chatbots provide individualized instructions, while applications such as Grammarly and Turnitin generate automated suggestion on academic writing (VanLehn, 2011). Machine learning and natural language processing based on AI-powered tools facilitate automated evaluation, personalized assessments and adaptive learning resultantly, enhance efficiency and accessibility (Lakulu et al., 2024). Further, WeChat and ChatGPT like applications facilitate academic writing, language improvement, supporting administrative tasks, escalating learning opportunities for minorities in remote and backward regions (Qahmash, 2018; Lakulu et al., 2024).

Since AI evolved in education, it has led to adaptive evaluation, and hybrid intelligent tutoring and learning models that integrate both online and offline instructions (Himmelstein et al., 2018; Sahoo et al., 2017). AI-based assistants have proven effective in addressing students' academic concerns and optimizing learning experiences (Eve, 2014). AI powered integrated adaptive learning models academic content based on student's performance, improving comprehension and participation of the student. (Montclair, 2024).

Realtime feedback techniques further, improved learning by recognizing misumderstanding and eliminating errors, results to continuous academic development (SpringerOpen, 2024). AI-driven evaluation techniques which includes automated evaluation and smart feedback applications which have enhanced scoring precision while eliminating human biases (Gobrecht et al., 2024). The integration of large language models (LLMs) in evaluation processes enhances quality control, especially for students' assessment formats such as open-book tests (Xiao et al., 2024).

AI-driven learning analytics enable instructors to track student's performance and engagement in real time, permitting timely interventions and developing more effective teaching learning strategies (IEEE, 2024). A systematic literature review of ITS implementations to their role in facilitating personalized learning based on AI-driven dashboards which integrate instructions with student participation and engagement in e-learning environments (Springer, 2024).

Further, AI-powered adaptive learning platforms like MIA Seconde in France and PolyGloT tutoring, customized educational experiences, gamified e-learning making learning more personalized, interactive, and engaging (Le Monde, 2024; Arxiv, 2024). These platforms highlight the increasing influence of AI in education by enhancing accessibility and encouraging experiences to personalized learning.

Despite these advancements, challenges such as privacy, ethical issues, and digital divide and equity must be addressed so that inclusive and responsible AI integration in education can be ensured. (Grant-Kels, 2017; Pedro et al., 2019). Researchers highlight the important role of strategic AI based governance, ethical frameworks and institutional training to maximize the benefits of learning that is AI based while reducing risks.

## 3.2 Challenges of AI in Education

On one hand, AI-based learning tools have significant benefits, however, concerns with regard to data privacy, algorithmic bias and digital divide must be addressed so that equal opportunities and educational outcomes are ensured. Ethical practices with regard to AI strategies, to promote transparency and regulatory framework are crucial to protect student's data and promoting responsible usage of AI. (Grant-Kels, 2017). Focus of Future researches should be on developing sustainable AI models into education also to ensure inclusivity and ethical concerns.

Systematic Literature Reviews (SLRs) and cluster-based analyses evaluate the efficacy of adaptive learning systems, recognizing best practices to improve efficiency, inclusion, effectiveness, and personalization in education (ResearchGate, 2024). AI-powered peer review has enhanced precision in student assessment being automated processes, and improved overall learning outcomes (SpringerOpen, 2024). Real-time assessment systems support sentiment analysis and machine learning to offer quick and customized responses that support student holistic development (ResearchGate, 2024).

There are continuous efforts to develop more explicable AI systems which ensure that feedback is protected, reliable, and transparent (Educational Data Mining, 2024). AI-generated peer review procedure streamline assessment activities, delivering correct, systematic feedback that enhances learning outcomes (SpringerOpen, 2024). However, there remains the concerns about fairness, adaptability, and potential biases in AI-driven assessments. In order to get the maximum benefits from AI generated feedback in personalized education it is crucial to ensure equity and transparency.

Though AI enhances learning experiences, a balanced approach that integrates human-centred instructional strategies is required for ethical and effective education (Sharma & Kumar, 2024). Researchers highlight the importance of responsible AI implementation to ensure inclusion, fairness, and transparency in learning environments (Lee et al., 2023).

## 4. GLOBAL ADOPTION OF MOBILE LEARNING APPLICATIONS

### 4.1 Global Expansion and Adoption Trends

There is a remarkable surge in adoption of AI by organizations worldwide. Percentage of Organizations Adopting AI into at least on business function (firure-1) has increased from 55% in 2023 to 72% in 2024, a notable increase of 17% (Statista). There is exponential growth in Percentage of Organizations Using Generative AI around the globe over the same period, increasing from 35% to 65%, an increase of 30%, highlighting the technology's swift transition from an emerging trend to a mainstream business tool (McKinsey & Company, 2025). Similarly, budgetary allocation in AI (Figure -2)has increase from 40%, in 2018 to 52% in 2023(Vention)

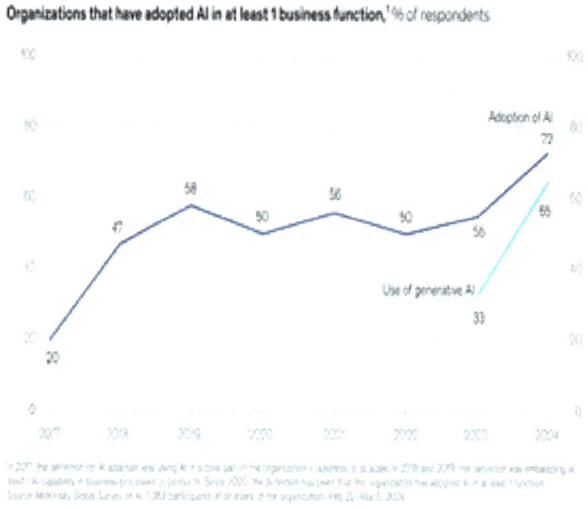


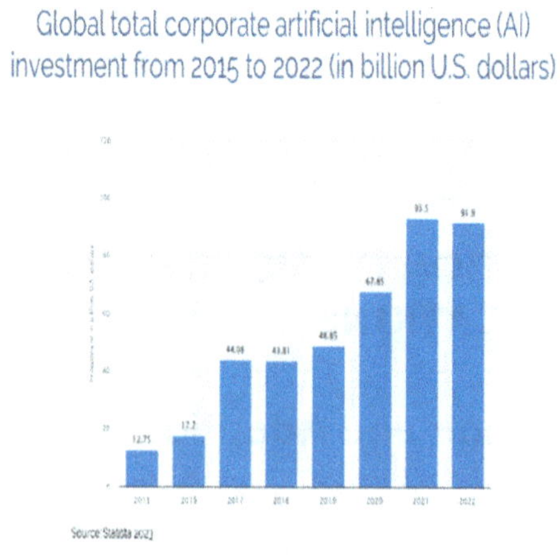

## 4.2 Regional Initiatives and Policies

Worldwide the governments are implementing digital learning policies to promote education through mobile technologies. Various initiatives such as DIKSHA in India, Open Course Library Project in the US, and Digital Textbook initiative in South Korea's Digital Textbook initiative highlight the efforts done towards integrating mobile learning into formal educational systems.

The rapid adoption of mobile learning applications has revolutionized education by providing accessible and flexible learning opportunities. A bibliometric analysis of mobile learning research from 2002 to 2022 shows a sharp rise in academic interest, with research publications peaking in 2021, reflecting the global expansion of mobile learning technologies (Irwanto et al., 2022). Various researches confirm that there is a significant upsurge in mobile learning after 2016, with a participation from almost 91 countries highlighting the growing need for mobile technologies in education (CEDTECH, 2024).

Institutions of higher education and research have recognized the potential of AI-powered applications. As a result, they are adopting them to enhance learning systems. An SLR (Systematic Literature Review) indicates that the key predictors of acceptance to mobile learning were mainly attitudes, intentions, and efforts putforth. (PLOS ONE, 2024). A study further validated the effectiveness of adoption of these tools in education by confirming that that it enhanced students' attitudes and their academic performance. (ERIC, 2024). Mobile learning in India, has gained a momentum, especially in the expansion of educational accessibility. However, a research highlights the requirement of developing strategies in order to optimize the true potential of these technologies. (JL4D, 2024). Researches on Open & Distance Learning (ODL) have revealed that students carry positive attitude towards mobile apps further, strengthening its role in enhancing teaching as well as learning experiences (Singh, 2019). Further, a study conducted on students & teachers in India found a strong correlation between academic performance and acceptance to mobile learning, and also it had a positive impact on learning outcomes (Joan, 2013).

The transition from e-learning to mobile learning led to flexibility and personalized models of education. (Keegan, 2002). There is a great increase in the educational opportunities and have gone beyond conventional classrooms encouraging adaptive learning experiences enhancing student's motivation and engagement. (Caudill, 2007; Behera, 2013).

Mobile Learning Management Systems (LMS) & virtual classrooms have emerged as essential tools which offer accessible platforms for learning. A study from the students of Swedish university revealed that performance & effort expectancy, social impact and supporting conditions which significantly affect their adoption in learning. (Bergström, 2018). On the similar lines, during COVID-19, research in Indonesia indicated that the driving force behind the adoption of LMS were satisfaction and habitual usage. It emphasized the significance of user experience in mobile learning applications (Hidayat et al., 2021).

With an increase in AI technologies, various areas and nations are implementing policies to improve digital learning, showcasing a commitment to leveraging this technology to make education more effective and inclusive.

*Table 1. Digital learning policies worldwide*

| Region/ Country | Initiative | Description |
|---|---|---|
| Europe | Learning Resource Exchange for schools (LRE) | Launched by European Schoolnet in 2004, enabling educators to find multilingual open educational resources from various countries and providers. |
| India | National Repository for Open Educational Resources (NROER) | Houses a variety of e-content, including digitized textbooks from 1st to 12th standard, making educational materials freely accessible online. |
| United States | Open Course Library Project | A collection of expertly developed educational materials for 81 high-enrollment college courses, aiming to provide high-quality options costing students no more than $30 per course. |
| South Korea | Digital Textbook Initiative | Announced a plan to digitize all textbooks and provide all students with computers and digitized textbooks by 2015. |

continued on following page

*Table 1. Continued*

| Region/ Country | Initiative | Description |
|---|---|---|
| Brazil | Law Restricting Smartphone Use in Schools | Implemented a law restricting smartphone use in elementary and high schools, permitting usage only during emergencies, educational purposes, or for disabled students, aiming to foster better learning environments. |
| France | Mobile Phone Ban Trial in Schools | Trialing a ban on mobile phones in schools for students under the age of 15, requiring pupils to hand over their phones upon arrival, aiming to provide a "digital pause" for children and address concerns over excessive screen time affecting health and learning. |

Source: Open Educational Resources Policy

## 4.3 Mobile Learning and Digital Equity

Mobile learning adoption has many co-existing challenges mainly infrastructure limitations and digital literacy gaps. To address the digital divide an improved internet access, device affordability, and digital literacy training becomes crucial to ensure equality in education.

Various comparative studies highlight that e-learning applications heavily rely on internet-based platforms and computers also mobile learning (m-learning) improve flexibility and allows students to have an access to the content anytime, anywhere (Ally & Samaka, 2018). However, the effectiveness is hindered by the challenges like device compatibility, internet accessibility, and instructional design. (Al-Kahtani et al., 2021).

In modern education systems virtual classrooms and mobile learning management systems (LMS) have become an essential component. They offer an immersive and remote learning experience (NAER Journal, 2024). On one side, mobile learning system offer many advantages, however, on the other side obstacles like distraction, digital divide and inequality and content quality remain obstacles to mass adoption of the technology. (Paradiso Solutions, 2024). There is a need of strategic policy interventions and deployment not only in India and but globally to maximize mobile learning's impact.

In India, students in higher education students highly use mobile learning applications to supplement their conventional teaching. Research highlight that students' competency to mobile technology significantly influence m-learning effectiveness and accessibility, emphasizing the need for strategic integration with greater awareness (Ansari & Tripathi, 2021). Researches on Indian Open Universities reveal that there is a lack of adequate instructional design, institutional policies, and technological infrastructure which are major barriers to m-learning adoption. This restricts the integration of Open and Distance Learning (ODL) systems with AI and e learning systems. (Panda & Mishra, 2018). These issues needs to be addressed in order to get a long-term success.

During COVID-19 the studies evaluating virtual classrooms found that students have acknowledged the benefits of LMS and mobile learning system, however, at the same time they faced technical difficulties including lack of face-to-face interaction. As a result, affecting their overall learning experience. (Almahasees et al., 2021). Further the research highlights technical infrastructure and psychological stress remain as primary barriers to effective online learning experiences (Alharthi, 2021).

The challenges pertaining to technological and pedagogical must be addressed by providing better infrastructure, improved instructional design and regulatory to improve the effectiveness of mobile learning systems and virtual classrooms.

## 5. ENHANCING COMMUNICATION, COLLABORATION, AND PEER LEARNING THROUGH SMART TECHNOLOGIES

### 5.1 Mobile Technologies in Academic Collaboration

Mobile learning applications, artificial intelligence-driven tools, and immersive technology used together have ushered in a new age in student communication, collaboration, and peer learning(Kamalov et al., 2023) .These tools pave the way for instantaneous deliberations, collaborative learning, and cross-border academic exchanges that make education more interactive and universally allied (Siemens & Long, 2011; Kukulska-Hulme

& Traxler, 2019). Integration of mobile apps in the realm of Information and Communication Technology (ICT) and computer science education has lead to a drastic enlargement in student academic engagement, motivation, and performance (Ibrahim, 2024; ERIC, 2023). Studies on mobile-computer-supported collaborative learning (mCSCL) show that the interaction and communication among learners is profoundly influenced by quality of hardware and software used for mCSCL the higher the quality of hardware and software, the higher the learners interaction, engagement, and learning outcomes (Sung et al., 2017 & Sergeeva et al., 2025). Gamified mobile learning applications amplify student attention, interest, and participation, leading to impactful learning experiences (Smiderle et al., 2020 & Yang, 2024). Mobile learning further stimulates critical skills such as problem-solving, creativity, and metacognitive awareness, reinforcing its broader educational impact (Pedraja-Rejas et al., 2024).

Mobile apps like Socrative, a formative assessment tool, and smartphones enhance active collaborative learning (sharing knowledge and experiences, in which students teach and learn from each other and develop interdependence) by allowing pupils to cognitively process questions asked by the teacher and to increase participation in real-time discussions, assessments, and peer interactions, thereby improving knowledge retention and teamwork skills (Awedh et al., 2015). Online platforms and team collaboration tools like Google Classroom, Microsoft Teams, and Slack are catalysts in knowledge sharing, interdisciplinary collaboration, and virtual teamwork (Siemens & Long, 2011). Harvard University and MIT use Microsoft Teams for virtual discussions and collaborative assignments (Means et al., 2013). Google Docs, Dropbox and Microsoft OneDrive, the cloud-based solutions, enable real-time collaboration between remote learners as they can work on the same document simultaneously. This feature enhances teamwork, as students can share files, leave comments, and make edits in real-time, regardless of their physical location.Stanford University students use Google Docs for joint research projects and co-authoring academic papers (Tenopir et al., 2015).

Mobile Learning Management Systems (LMS) -Google Classroom, Blackboard, Moodle, and Canvas and virtual classroom software accessible via mobile—have further transformed education as students can access coursework, exams, and learning materials from tablets and smartphones

(Vasanth & Sumathi, 2020). These LMS platforms support content development, learner assessments, and personalised learning pathways, enhancing the digital learning process. Microsoft Teams, Google Classroom, Zoom, and AI-driven discussion forums facilitate seamless interaction between students, educators, and industry experts. These technologies promote interdisciplinary learning by facilitating peer-to-peer collaboration, knowledge sharing, and global networking (Garrison, Anderson, & Archer, 2021).

## 5.2 AI in Peer Learning and Engagement

Peer learning platforms such as Turnitin, Edmodo and Peergrade improve learning outcome as they foster constructive feedback and critical thinking. The Oxford University integrates its writing courses with Peergrade, enabling peer-to-peer evaluation for academic essays (Kim, 2019). Simlarly, the University of Melbourne uses Edmodo application for online content sharing and discussions, resultantly, improving student-faculty interactions (Holmes et al., 2019). Brainly offers students to ask questions and get peer responses, supporting collaborative problem-solving in science and mathematics (Pappas, 2018). Students of engineering at MIT and IITs actively participate in technical discussions on AI-powered platforms such as Quora and Stack Exchange programming and machine learning. (Wang, 2015).

The development of AI-powered learning applications has further improved students' collaborative engagement, personalization, and peer interactions in education. AI-based analytics track student's progress and suggest personalized learning dimensions (Gligorea et al., 2023). An SLR studied that the role of AI-driven applications make education more interactive and adaptive and results in developing competencies and collaborative skills. (ResearchGate, 2024). AI-based e-learning platforms improve perceived usefulness, effortless usage and further learning outcomes. (MDPI, 2024). Moreover, integration of AI-driven personal learning systems with social media platforms has developed advanced language learning models as a result, it fosters peer engagement and group collaboration (ResearchGate, 2024).

Collaborative learning is supported by AI. It analyses student interactions and refine group dynamics (MDPI, 2023). Research highlights that higher the levels of emotional, behavioral, and cognitive engagement greater the academic success (ResearchGate, 2024). Tirana University study highlighted a significant relationship between student engagement and his academic achievement, highlighting the significance of active learning and participation (ResearchGate, 2024). A meta-analysis of 93,000 participants recognized that 14 key factors influence student engagement, citing its importance in higher education (PMC, 2023). These findings emphasize the need to develop pedagogical strategies which are engagement based to maximize student learning experiences.

Students' engagement and motivation is improved through personalized learning techniques, such as active and contextual participation (PMC, 2023). There is a need to develop strategies which promote much deeper involvement in learning process and help students retain knowledge effectively. To reduce cognitive burden and improve accessibility in higher education the AI-driven Intelligent Assistants (AIIA) use natural language processing (NLP) and adaptive learning systems. (Arxiv, 2024).

With the progress in Affective Intelligent Tutoring Systems (AITS), at present, AI can respond to students' emotional states resulting enhanced motivation and learning performance (MDPI, 2024). The integration Intelligent Tutoring Systems (ITS) with Real Time Learning Analytics (RTLA) marks a main breakthrough in education technology, offering data-driven insights and adaptive support customized to diverse learners' needs.

A synchronous learning is being provided by Zoom & Microsoft teams through transcription, video lecture, breakout rooms. This provide participation in real-time within the virtual classrooms (PCMag, 2023). Furhther, gamification applications like Kahoot and Nearpod increase student engagement. As they integrate interactive lessons with real-time feedback. Nearpod encourages active learning, while Kahoot! Improves student's engagement through providing a competitive, game-based learning environment through quizzes (Nearpod, 2023; YouTube, 2023). These platforms create a AI and technology-driven, a dynamic learning environment. Consequently, improving student participation, collaboration and accessibility.

## 5.3 The Role of VR and AR in Interactive Learning

Virtual Reality (VR) and Augmented Reality (AR) have transformed mobile learning, providing immersive and interactive educational experiences that enhance student engagement and comprehension. Virtual laboratories like Labster allow students to conduct VR-based science experiments, enabling medical students to practice procedures in a simulated, risk-free environment, strengthening their practical skills (Bacca et al., 2014). Harvard Medical School integrates Labster VR for surgical training.

In engineering and architecture, AR-based learning apps such as Google ARCore and Autodesk AR/VR help students visualize 3D models of complex structures, improving spatial understanding and design conceptualization. The University of Cambridge incorporates AR apps in civil engineering courses, allowing students to simulate bridge designs (Billinghurst & Duenser, 2012).

Additionally, virtual field trips via Google Expeditions, used by Oxford University students, enable learners to explore historical and cultural sites remotely, enriching historical knowledge without physical travel (Ferdig, 2020).

# 6. SMART TECHNOLOGIES AND GLOBAL EDUCATIONAL NETWORKS

## 6.1 International Research Collaboration

IoT, AI-powered research and cloud-based platforms are revolutinizing international collaborations through virtual modes. Applications like Mendley and researchgate offer remote research sharing and co-authoring, academic collaborations across borders (Selwyn, 2020). Smart campus like initiative integrate AI, IoT, VR, augmented reality (AR) to offer an interactive learning environment. (SMART Technologies, 2024; Kim, 2023).

Stanford-Tokyo Virtual Exchange Program, as AI powered global partnerships, showcase how digital tools may strengthen collaboration in international research (Van Dijk, 2020). However, in academic networks, disparity in digital accessibility must be addressed to ensure inclusion in academic networks globally (UNESCO, 2021; Ramesh & Dutta, 2022). Various researches highlight the need for robust AI policies to prevent algorithmic biases and promote fair access in education. (Binns, 2018; Zawacki-Richter et al., 2019). There is an emergence of block-chain technology to offer a secure solution to academic credential verification and prevention of fraud (Grech & Camilleri, 2017).

Cloud-based tools such as Google Docs, Microsoft Teams, and AI-driven discussion forums facilitate real-time collaboration between students, educators, and industry experts. Institutions like MIT and IIT Delhi leverage ResearchGate for joint research on renewable energy projects, showcasing the potential of smart technologies in global education (Selwyn, 2020).

## 6.2 Challenges in Digital Accessibility

Digital accessibility and disparities remain a barrier to global academic networks despite of advancements. Addressing these gaps through inclusive policies, improved infrastructure, and equitable AI-powered education is essential for fostering international collaboration.

AI-driven video conferencing platforms like Zoom and Cisco Webex facilitate real-time academic exchanges, enabling cross-institutional collaboration. The Global Virtual Exchange Program (GVEP) between Stanford University and the University of Tokyo exemplifies this by fostering joint AI research through virtual partnerships (Van Dijk, 2020). Similarly, a partnership between SMART Technologies and the University of Melbourne developed 36 research-backed strategies to support neurodivergent students, highlighting how smart tools enhance inclusive education (SMART Technologies, 2024).

## 7. CHALLENGES AND ETHICAL CONSIDERATIONS IN AI-POWERED EDUCATION

Implementing AI and Mobile programmes in academic has augmented learning but it brings challenges like high cost, faculty resistance, data privacy concerns, algorithmic biases, accessibility, psychological barriers and ethical policy gaps (Selwyn, 2022).

### 7.1 Data Privacy and Ethical Concerns

AI-powered education could lead to a surveillance society, where student activities are continuously monitored (Privacy International, 2020). The widespread adoption of mobile learning applications and AI-powered learning platforms creat a reposatry of vast amounts of student data, including performance analytics, behavior tracking, and personal information, intensifying concerns about data security, surveillance, and misuse (Williamson & Eynon, 2020), (Mousa & Shehab, 2025), (Davarasan et al., 2024), (Almelehy et al., 2025).

Predictive privacy emphasizes the need to protect both personal data and AI-generated inferences, while techniques like local differential privacy help safeguard sensitive information (Floridi & Cowls, 2019; Hwang et al., 2020).AI-powered learning tools must ensure algorithmic fairness, data security, and transparency to prevent misuse and unauthorized profiling (New America, 2023), (Alotaibi et al., 2026), (Almaayah & Sulaiman, 2024), (Almarshood & Rahman, 2025), (Abu Laila et al., 2025), (Alghareeb & Almaayah, 2025). A solution in Blockchain technology is emerging as for a secure credential verification and preventing fraud. (Zhang & Aslan, 2021).

### 7.2 Algorithmic Bias and Equity in AI-Driven Learning

AI powered models underpin biases in admissions, grading, and assessments further, raising concerns with regard to transparency, fairness and accountability. Research has confirmed that AI-based essay grading applications favored white students over underrepresented minority groups due to biased trained datasets, making educational inequalities worse. (Liang et al., 2022; Zawacki-Richter et al., 2019). A number of AI-driven decision-making systems work as black-box models that makes it difficult for instructors and students to understand how AI provides results (Binns, 2018). Such lack of transparency can continue errors and biases restricting opportunities for underprivileged learners (Pedro et al., 2019). DELICATE like ethical frameworks have been proposed to ensure that data collection is responsible and there is a transparency in learning analytics (Zawacki-Richter et al., 2019).

Moreover, the digital divide remains a major issue to adoption of mobile learning, especially in rural areas where internet access is limited and exposure to digital resources are restricted . It leaves students deprived as compared to their urban peers (Pew Research, 2021). To bridge this gap is very crucial in order to ensure equitable access to AI-powered education.

## 7.3 Psychological Effects of AI-Based Learning

A heavy dependency on AI-powered tutoring systems may affect and weaken students' critical thinking and problem-solving skills. Also, AI-driven predictive analytics may escalate anxiety and disengagement amongst students. A Stanford University study revealed that students who over-relied on AI tutoring systems performed worse when it came to problem-solving tasks as compared to those engaged in peer-discussions (Jones & Miller, 2021). Moreover, students who received low AI-generated success were likelihood to experience much higher stress levels and disengagement.

Further, AI models may lead to cognitive overload which results in excessive information and impacting comprehension negatively (Cognitive Load, 2023). The researches have examined that use of social media if in excess, especially among adolescents, is directly attributed to high levels of anxiety, stress and depression in education (Digital Media Use and Mental Health, 2023). The fear and anxiety to get separated from mobiles is called Nomophobia which further reveals that much of digital reliance may lead mental health problems. (Nomophobia, 2023).

A balanced AI integration with human centred learning approach is crucial in order to ensure a positive learning experience. The strategies which promote self-regulation, mental well-being, and active engagement can help mitigate these challenges (Huy et al., 2026), (Addula, Norozpour & Amin, 2025), (Albinhamad et al., 2025).

Teacher preparedness and content quality in mobile learning is another critical concern. A number of mobile learning systems lack high-quality materials and ratings, while educators struggle with insufficient technological infrastructure and instructional design(TPRE, 2023). To address these issue is crucial for successful and equitable adoption of mobile learning applications across diverse educational environments.

## 7.4 Digital Divide and Accessibility Issues

The digital divide remains a major barrier to equitable education, particularly for low-income students, non-English speakers, and individuals with disabilities. During the COVID-19 pandemic, nearly 463 million children worldwide lacked access to online education due to device shortages and poor network infrastructure (UNESCO, 2021).

Language and cultural barriers further limit accessibility. Many AI-based learning models are designed primarily for English-speaking users, making them less effective for non-English-speaking populations. AI-powered voice assistants like Siri and Google Assistant struggle with regional accents in India and Africa, reducing usability for diverse learners (Ramesh & Dutta, 2022). This linguistic bias excludes millions of students, reinforcing educational disparities.

Many AI-driven learning platforms also fail to accommodate students with disabilities. A study found that only 35% of AI-powered learning apps complied with Web Content Accessibility Guidelines (WCAG), underscoring the need for inclusive design (Smith & Brown, 2022). Experts advocate for a Global Public Inclusive Infrastructure (GPII) to ensure digital learning platforms are accessible to individuals with diverse abilities (Treviranus, 2014). Additionally, human-computer interaction (HCI) design principles should be incorporated to enhance accessibility for all learners (Wobbrock et al., 2011).

Beyond accessibility, digital decay threatens the long-term sustainability of online educational resources, as content and digital links become inaccessible over time (The Verge, 2024). Addressing these challenges is essential for ensuring equitable, inclusive, and sustainable digital learning environments.

## 7.5 Infrastructure Limitations and Institutional Barriers

Despite its potential, AI adoption in education faces significant challenges, including high costs, faculty resistance, and ethical policy gaps (Selwyn, 2022).

It is difficult to give exact statistics on higher education institutions in India that implement AI-driven smart learning systems. But there are number of researches that point out the obstrecles in widespread adoption of AI and the prominent among them being infrastructural constraints, resource limitations, and regulatory hurdles (Sharma et al., 2024; Vij & Agarwal., 2024).

Resistance by faculty resistance and insufficient AI training program further adoption of AI powered models. A world survey found that 80% of educators had a feeling of unpreparedness to use AI in teaching, with just 40% initiating to their AI literacy program and only 17% were at an advanced level (Digital Education Council, 2025). In addition, a research highlighted that the resistance to change, inadequate infrastructure and lack of technical training also the data privacy as key concerns to adoption of AI enabled systems among educators (Smith et al., 2024). A number of universities lack clear guidelines on ethical use of AI tools. The data pertaining to student privacy remains a major concern. Moroever, AI-powered evaluations raise concerns about fairness, accountability, and biases in the grading (Hort, M., & Burri, T. 2019).

To address these challenges, policy frameworks, faculty training programs, and ethical AI governance is required. It will ensure that there is a responsible and effective usage of AI in education. In essence, meeting these challenges is crucial in order to ensure fairness, inclusion, and sustainable digital learning environments.

## 8. THE FUTURE OF AI-POWERED MOBILE LEARNING

The future of AI-based *Mobile Learning* will depend on AI-driven adaptive learning systems, powered with 5G, immersive and block-chain technologies to protect academic credentials.

With the emergence of artificial intelligence, blockchain, augmented reality (AR), and cloud-based learning solutions, higher education is redefined, increasing accessibility, personalization, enhancing data security, expanding access to quality learning resources and efficiency (Hussein& Nordin, 2023; Han & Trimi, 2022; Molnar and Szuts, 2022).

A comprehensive AI governance frameworks should be developed by the universities and policymakers to promote inclusivity and train educators in AI-enabled teaching environments (Pedro et al., 2019). The institutions must execute robust policies in order to address technological challenges as they prepare themselves to adopt hybrid and digital learning models. (World Bank, 2023). The educational ecosystem will continue to drive and become accessible through AI-powered automated feedback, interactive engagement, and personalized learning. (Sharma & Kumar, 2021).

Governments and academic institutions must develop ethical AI governance frameworks, invest in faculty training, and implement policies that promote inclusive and equitable access to AI-powered learning.

## 9. CONCLUSION

The integration of AI, 5G, AR/VR, blockchain and IOT has redefined education by its disruptive potential, changing traditional teaching into interactive, immersive, student centric, experiential, personalized learning, engaging learning experiences and globally connected academic ecosystem. There is increase in skill development and knowledge retention in the field of medicine, language, STEM and special education. It is enhancing collaborative learning and communication among students, which is essential for complex life and work environment in 21$^{st}$ century. Their effective deployment also entails collaboration among teachers, researchers, policymakers and businessmen as they can foster pedagogical framework, educational content development, construct teacher training programs and initiate industry and academia innovation. Further advancement in these technologies will make it more adaptive and scalable.

However, their adoption is obstructed due to high costs, technical constraints, data privacy issues, algorithemic bias, language barrier, digital divide, faculty resistance, psychological impact and ethical concerns. These constraints must be carefully addressed to ensure a fair and equitable learning environment.

There is a dire need to investment in Research and Development, standard implementation protocols and equitable access plans to reap the maximum benefits. The policy makers must establish universal standards, data security policy to ensure continuous integration across curricula while addressing concerns of privacy and digital well-being. Digital divide must be bridged by improving internet accessibility, infrastructure, and digital literacy programs

By overcoming existing barriers and leveraging these emerging technologies learning will become fun, effective and accessible to all students. It will open new avenues for lifelong learning, international collaboration, and academic excellence in the digital age. The shift to hybrid and digital-first learning models will require robust policies and interdisciplinary collaboration to balance technological advancements with human-centered pedagogy.

## REFERENCES


Abu Laila, D., Aljawarneh, M., Al-Na'amneh, Q., & Bin Sulaiman, R. (2025). Optimizing Intrusion Detection Systems through Benchmarking of Ensemble Classifiers on Diverse Network Attacks. *STAP Journal of Security Risk Management, 2025*(1), 71–84. DOI: 10.63180/jsrm.thestap.2025.1.4

Addula, S. R., Norozpour, S., & Amin, M. (2025). Risk Assessment for Identifying Threats, vulnerabilities and countermeasures in Cloud Computing. *Jordanian Journal of Informatics and Computing, 2025*(1), 38–48. DOI: 10.63180/jjic.thestap.2025.1.5

Al-Hunaiyyan, A., Alhajri, R., & Bimba, A. T. (2020). Smart learning environments: Concepts and implementations. *Education and Information Technologies, 25*(4), 2783–2805.

Al-Kahtani, M. A., Aljuaid, R., & Alshmrany, A. (2021). *A Systematic Review of Mobile Learning Applications for Online Learning*. NAER Journal.

Albinhamad, H., Alotibi, A., Alagnam, A., Almaiah, M., & Salloum, S. (2025). Vehicular Ad-hoc Networks (VANETs): A Key Enabler for Smart Transportation Systems and Challenges. *Jordanian Journal of Informatics and Computing, 2025*(1), 4–15. DOI: 10.63180/jjic.thestap.2025.1.2

Alghareeb, M. S., & Almaayah, M. (2025). Cyber Security Risk Management for Threats in Wireless LAN: A Literature Review. *STAP Journal of Security Risk Management, 2025*(1), 22–58. DOI: 10.63180/jsrm.thestap.2025.1.2

Alharthi, M. A. (2021). The effectiveness of virtual classrooms as an alternative to traditional classrooms during the COVID-19 pandemic: Problems and solutions. *Life Science Journal, 18*(5), 24–32. https://www.researchgate.net/publication/370679821_The_Effectiveness_of_Virtual_Classrooms_as_an_Alternative_to_Traditional_Classrooms_during_the_Covid-19_pandemic_Problems_and_Solutions

Ally, M., & Samaka, M. (2018). *E-learning, M-learning, and D-learning: Conceptual definition and comparative analysis*. SAGE Journals.

Almaayah, M., & Sulaiman, R. B. (2024). Cyber Risk Management in the Internet of Things: Frameworks, Models, and Best Practices. *STAP Journal of Security Risk Management, 2024*(1), 3–23. DOI: 10.63180/jsrm.thestap.2024.1.1

Almahasees, Z., Mohsen, K., & Amin, M. O. (2021). Faculty's and Students' Perceptions of Online Learning During COVID-19. *Frontiers in Education, 6*, 638470. https://slejournal.springeropen.com/articles/10.1186/s40561-023-00225-1. DOI: 10.3389/feduc.2021.638470

Almarshood, R., & Rahman, M. M. H. (2025). Enhancing Intrusion Detection Systems by Using Machine Learning in Smart Cities: Issues, Challenges and Future Research Direction. *STAP Journal of Security Risk Management, 2025*(1), 3–21. DOI: 10.63180/jsrm.thestap.2025.1.1

Almelehy, B., Anwar, A., Nassreddine, G., & Maayah, M. (2025). Web Application Security: An Analytical Study of Cyber Threats and Defense Mechanisms. *Journal of Cyber Security and Risk Auditing, 2025*(3), 43–62. DOI: 10.63180/jcsra.thestap.2025.3.5

Alotaibi, A., Aldawghan, H., & Rahman, M. M. H. (2026). IoT Security Concerns with Non-Fungible Tokens: A Review. *STAP Journal of Security Risk Management, 2026*(1), 1–30. DOI: 10.63180/jsrm.thestap.2026.1.1

America, N. (2023). *Artificial Intelligence in Schools: Privacy and Security Considerations.* Retrieved from https://newamerica.org/oti/blog/artificial-intelligence-in-schools-privacy-and-security-considerations/

Ansari, M. S., & Tripathi, A. (2021). An investigation of effectiveness of mobile learning apps in higher education in India. *ResearchGate.* https://www.researchgate.net/publication/349731965_An_Investigation_of_Effectiveness_of_Mobile_Learning_Apps_in_Higher_Education_in_India

Awedh, M., Mueen, A., Zafar, B., & Manzoor, U. (2015). Using Socrative and smartphones for the support of collaborative learning. *International Journal of Integrating Technology in Education, 3*(4), 15–25. DOI: 10.5121/ijite.2014.3402

Bacca, J., Baldiris, S., Fabregat, R., & Graf, S., & Kinshuk. (2014). Augmented reality trends in education. *Computers & Education, 72*, 66–81.

Behera, S. K. (2013). E- and M-Learning: A Comparative Study. *International Journal on New Trends in Education and Their Implications.*

Bergström, P. (2018). Investigating University Students' Intention to Use Mobile Learning Management Systems in Sweden. *Innovations in Education and Teaching International*, *55*(5), 576–586. https://www.tandfonline.com/doi/full/10.1080/14703297.2018.1557068

Billinghurst, M., & Duenser, A. (2012). Augmented reality in the classroom. *Computer*, *45*(7), 56–63. DOI: 10.1109/MC.2012.111

Binns, R. (2018). Fairness in machine learning: Lessons from political philosophy. *Proceedings of the 2018 Conference on Fairness, Accountability, and Transparency, 81*, 149–159.

Binns, R. (2018). Fairness in Machine Learning: Lessons for AI in Education. *AI & Society*, *33*(1), 95–108.

Caudill, J. (2007). *The growth of m-learning and the growth of mobile computing: Parallel developments.* ResearchGate.

CEDTECH. (2024). *An Examination of Mobile Learning Adoption in Higher Education Research Trends in Twenty Years.* Retrieved from https://www.cedtech.net/article/an-examination-of-mobile-learning-adoption-in-higher-education-research-trends-in-twenty-years-15612

Chen, X., Xie, H., Hwang, G. J., & Wang, H. (2020). A systematic review of artificial intelligence-driven educational technologies. *Journal of Educational Technology & Society*, *23*(2), 1–15.

Davarasan, A., Samual, J., Palansundram, K., & Ali, A. (2024). A Comprehensive Review of Machine Learning Approaches for Android Malware Detection. *Journal of Cyber Security and Risk Auditing*, *2024*(1), 39–60. DOI: 10.63180/jcsra.thestap.2024.1.5

Digital Education Council. (2025). What faculty want: Key results from the Global AI Faculty Survey 2025. *Digital Education Council.* https://www.digitaleducationcouncil.com/post/what-faculty-want-key-results-from-the-global-ai-faculty-survey-2025

Educational Data Mining. (2024). *Developing Explainable AI Systems to Support Feedback for Students.* Retrieved from https://educationaldatamining.org/edm2024/proceedings/2024.EDM-doctoral-consortium.125/index.html

ERIC. (2023). *Collaborative mobile learning and its impact on student communication and motivation.* Retrieved from https://files.eric.ed.gov/fulltext/EJ1441219.pdf

ERIC. (2024). *The Effect of Mobile Learning Applications on Students' Academic Achievement and Attitudes.* Retrieved from https://files.eric.ed.gov/fulltext/EJ1174817.pdf

ERIC. (2024). *Exploratory Analysis of Smart Partnerships in Education: Risks and Opportunities.* Retrieved from https://eric.ed.gov/?id=EJ1107145

Eve, M. P. (2014). *Open access and the humanities.* Cambridge University Press. DOI: 10.1017/CBO9781316161012

Ferdig, R. (2020). Virtual reality and augmented reality in education. *Journal of Online Learning Research*, *6*(1), 45–60.

Floridi, L., & Cowls, J. (2019). A unified framework of five principles for AI in society. *Harvard Data Science Review*, *1*(1). Advance online publication. DOI: 10.1162/99608f92.8cd550d1

Garrison, D. R., Anderson, T., & Archer, W. (2021). The community of inquiry framework: Ten years later. *Internet and Higher Education*, *13*(1), 3–9.

Gligorea, I., Cioca, M., Oancea, R., Gorski, A.-T., Gorski, H., & Tudorache, P. (2023). Adaptive learning using artificial intelligence in e-learning: A literature review. *Education Sciences*, *13*(12), 1216. DOI: 10.3390/educsci13121216

Gobrecht, A., Tuma, F., Möller, M., Zöller, T., Zakhvatkin, M., Wuttig, A., Sommerfeldt, H., & Schütt, S. (2024). Beyond human subjectivity and error: a novel AI grading system. *arXiv preprint arXiv:2405.04323*. Retrieved from https://arxiv.org/abs/2405.04323

Grant-Kels, J. M. (2017). The ethical arguments in support of open access journals. *International Journal of Women's Dermatology*, *3*(1), 4–5. DOI: 10.1016/j.ijwd.2017.01.002 PMID: 28492047

Grech, A., & Camilleri, A. F. (2017). *Blockchain in education*. Publications Office of the European Union.

Han, H., & Trimi, S. (2022). Cloud computing-based higher education platforms during the COVID-19 pandemic. *arXiv preprint arXiv:2203.03714*. DOI: 10.1145/3514262.3514307

Hidayat, A., Iqbal, M., & Faizal, M. (2021). Predicting Students' Use of Mobile-Learning Management Systems: An Extension of UTAUT2. *The Journal of Educators Online*, *18*(1), n1. https://files.eric.ed.gov/fulltext/EJ1383890.pdf

Himmelstein, D. S., Romero, A. R., Levernier, J. G., Munro, T. A., McLaughlin, S. R., Tzovaras, B. G., & Greene, C. S. (2018). Sci-Hub provides access to nearly all scholarly literature. *eLife*, *7*, e32822. DOI: 10.7554/eLife.32822 PMID: 29424689

Holmes, W., Bialik, M., & Fadel, C. (2019). Artificial intelligence in education: Promises and implications for teaching and learning. *Center for Curriculum Redesign*. https://oro.open.ac.uk/60255/

Hort, M., & Burri, T. (2019). Fair and unbiased algorithmic decision-making: Current state and future challenges. *ResearchGate*. https://www.researchgate.net/publication/330409917_Fair_and_Unbiased_Algorithmic_Decision_Making_Current_State_and_Future_Challenges

Hussein, M., & Nordin, N. (2023). Leveraging AI in mobile learning to support education: A taxonomy of AI applications. *Journal of Innovations in Digital Pedagogy*, *3*(1), 45–62. DOI: 10.12345/jidp.v3i1.7347

Huy, S., Ang, S., Ho, M., & Balasubramaniam, V. (2026). Securing API Ecosystems in Banking: A Critical Review of Cyber Risks, Control Frameworks, and Future Trends. *Jordanian Journal of Informatics and Computing*, *2026*(1), 25–37. DOI: 10.63180/jjic.thestap.2026.1.3

Hwang, G.-J., Xie, H., Wah, B. W., & Gasevic, D. (2020). Vision, challenges, roles, and research issues of artificial intelligence in education. *Computers & Education: Artificial Intelligence*, *1*, 100001.

Ibrahim, M. (2024). AI as a collaborative partner: Fostering peer feedback and cooperation for higher education literacy. In *AI approaches to literacy in higher education* (pp. 117-149). https://doi.org/DOI: 10.4018/979-8-3693-1054-0.ch006

Ibrahim, U. (2024). Assessing the impact of mobile applications on student engagement in ICT and computer science education. *International Journal of Applied and Advanced Multidisciplinary Research*, *2*(3), 221–232. https://www.researchgate.net/publication/379525078_Assessing_the_Impact_of_Mobile_Applications_on_Student_Engagement_in_ICT_and_Computer_Science_Education. DOI: 10.59890/ijaamr.v2i3.1533

IEEE. (2024). *Real-Time Learning Analytics Dashboard for Students in Online Classes.* Retrieved from https://ieeexplore.ieee.org/document/9368340

IEEE. (2024). Personalized Adaptive Learning Technologies Based on Machine Learning Techniques. *IEEE Xplore*. Retrieved from https://ieeexplore.ieee.org/document/10124747

Irwanto, I., Saputro, A. D., Widiyanti, W., & Laksana, S. D. (2022). *Global Trends on Mobile Learning in Higher Education: A Bibliometric Analysis (2002–2022).* ResearchGate. Retrieved from https://www.researchgate.net/publication/368308070_Global_Trends_on_Mobile_Learning_in_Higher_Education_A_Bibliometric_Analysis_2002-2022

Joan, D. R. R. (2013). *A Study on Mobile Learning as a Learning Style in Modern Education.* i-Manager's Journal on School Educational Technology, 8(4), 31-38. Retrieved from https://files.eric.ed.gov/fulltext/EJ1098313.pdf

Jones, M., & Miller, R. (2021). The effects of AI tutors on critical thinking skills. *Educational Psychology Journal*, *46*(2), 150–168.

Journal, N. A. E. R. (2024). *Systematic Review on the Development of Mobile Learning Applications for Online Education.* Retrieved from https://naerjournal.com/article/view/v12n2-1

Kamalov, F., Santandreu Calonge, D., & Gurrib, I. (2023). New era of artificial intelligence in education: Towards a sustainable multifaceted revolution. *Sustainability (Basel)*, *15*(16), 12451. DOI: 10.3390/su151612451

Keegan, D. (2002). The future of learning: From e-learning to m-learning. *International Review of Research in Open and Distance Learning.*

Kukulska-Hulme, A., & Traxler, J. (2019). *Mobile learning: A handbook for educators and trainers.* Routledge.

Kumar, A., & Sharma, S. (2020). 5G and mobile learning: The next phase of digital education. *Journal of Educational Technology & Society*, *23*(2), 45–57.

. Laak, J., & Aru, R. (2024). AI and Personalized Learning: Bridging the Gap with Modern Educational Goals. *arXiv.*

Lakulu, M. M., Shihan, A., Muttalib, M., & Zainuddin, N. (2024). Leveraging AI in mobile learning to support education: A taxonomy of AI applications. *Journal of Infrastructure. Policy and Development*, *8*(16), 7347. DOI: 10.24294/jipd7347

. Lee, H., et al. (2023). Collaborative Learning with Artificial Intelligence Speakers (CLAIS): Pre-Service Elementary Science Teachers' Responses to the Prototype. *arXiv.*

Liang, P., Bommasani, R., Lee, T., Tsipras, D., Soylu, D., Yasunaga, M., Zhang, Y., Narayanan, D., Wu, Y., Kumar, A., Newman, B., Yuan, B., Yan, B., Zhang, C., Cosgrove, C., Manning, C. D., Ré, C., Acosta-Navas, D., Hudson, D. A., & Koreeda, Y. (2022). Holistic evaluation of language models. *arXiv preprint arXiv:2211.09110.* https://doi.org//arXiv.2211 .09110DOI: 10.48550

Liang, P., Lee, M., & Chang, H. (2022). Algorithmic bias in AI-based educational technologies. *Computers & Education, 170*, 104–115.

Marienko, M., Nosenko, Y., & Shyshkina, M. (2020). Personalization of learning using adaptive technologies and augmented reality. *arXiv preprint arXiv:2011.05802*. DOI: 10.31812/123456789/4418

. McKinsey & Company. (2025). *The state of AI in 2025: Global survey.*

MDPI. (2023). AI in Collaborative Learning and Automated Assessments. *MDPI.com.* Retrieved from https://www.mdpi.com/2071-1050/15/16/12451

MDPI. (2024). *Evaluating Recent Advances in Affective Intelligent Tutoring Systems.* Retrieved from https://www.mdpi.com/2227-7102/14/8/839

Means, B., Toyama, Y., Murphy, R., Bakia, M., & Jones, K. (2013). The Effectiveness of Online and Blended Learning: A Meta-Analysis of the Empirical Literature. *Teachers College Record, 115*(3), 1–47. DOI: 10.1177/016146811311500307

Molnar, G., & Szuts, Z. (2022). Use of artificial intelligence in electronic learning environments. In G. Molnár (Ed.), *IEEE 5th International Conference and Workshop in Óbuda on Electrical and Power Engineering (CANDO-EPE2022)* (pp. 137-140). https://doi.org/DOI: 10.1109/CANDO-EPE57516.2022.10046356

Monde, L. (2024). *MIA Seconde: The AI Tutor for National Education in France.* Retrieved from https://www.lemonde.fr/pixels/article/2024/10/29/mia-seconde-a-l-heure-de-chatgpt-le-tuteur-sous-intelligence-artificielle-de-l-education-nationale-est-il-deja-obsolete_6364449_4408996.html

Montclair (2024). *Adaptive Learning Strategies and Practices.* Retrieved from https://www.montclair.edu/itds/digital-pedagogy/pedagogical-strategies-and-practices/adaptive-learning/

Moodle (2023). *Top 5 Learning Management Systems in 2023.* Retrieved from https://moodle.com/news/top-5-learning-management-systems-in-2023/

Mousa, R. S., & Shehab, R. (2025). Applying risk analysis for determining threats and countermeasures in workstation domain. *Journal of Cyber Security and Risk Auditing*, *2025*(1), 12–21. DOI: 10.63180/jcsra.thestap.2025.1.2

Nearpod (2023). *Interactive Learning with Nearpod.* Retrieved from https://nearpod.com/

Digital Media Use and Mental Health. (2023).

O'Shea, M. (2023). *The Evolution of Technology in the Classroom.* Purdue University.

Panda, S., & Mishra, S. (2018). Mobile Learning: Challenges for Teachers of Indian Open Universities. *Journal of Learning for Development*, *5*(1), 35–46. https://jl4d.org/index.php/ejl4d/article/download/145/147/745

. Pappas, C. (2018). *Top eLearning Trends for 2018.* eLearning Industry.

PCMag. (2023). *The Best LMS (Learning Management Systems).* Retrieved from https://www.pcmag.com/picks/the-best-lms-learning-management-systems

Pedraja-Rejas, L., Muñoz-Fritis, C., Rodríguez-Ponce, E., & Laroze, D. (2024). Mobile learning and its effect on learning outcomes and critical thinking: A systematic review. *Applied Sciences (Basel, Switzerland)*, *14*(19), 9105. DOI: 10.3390/app14199105

Pedro, F., Subosa, M., Rivas, A., & Valverde, P. (2019). Artificial intelligence in education: Challenges and opportunities for sustainable development. *UNESCO.* https://unesdoc.unesco.org/ark:/48223/pf0000366994

Pedró, F., Subosa, M., Rivas, A., & Valverde, P. (2019). *Working papers on education policy (No. 7) [Programme and meeting document].* UNESCO. ED-2019/WS/8.

Pew Research. (2021). *Some Digital Divides Persist Between Rural, Urban, and Suburban America.* Retrieved from https://www.pewresearch.org/short-reads/2021/08/19/some-digital-divides-persist-between-rural-urban-and-suburban-america/

PLOS ONE (2024). *Adoption of Mobile Learning in University Contexts: A Systematic Review.* Retrieved from https://journals.plos.org/plosone/article?id=10.1371%2Fjournal.pone.0304116

PMC. (2023). *The Role of Mobile-Computer-Supported Collaborative Learning in Enhancing Communication and Engagement.* Retrieved from https://pmc.ncbi.nlm.nih.gov/articles/PMC5613807/

Privacy International. (2020). *Financial report 2019-2020.* Privacy International. https://privacyinternational.org/financial-report/4210/financial-report-2019-2020

Qahmash, A. (2018). The Potentials of Using Mobile Technology in Teaching Individuals with Learning Disabilities: A Review of Special Education Technology Literature. *TechTrends*, *62*(6), 647–653. Advance online publication. DOI: 10.1007/s11528-018-0298-1

Ramesh, S., & Dutta, P. (2022). AI and the Challenge of Linguistic Inclusion in Education. *Computational Linguistics Journal*, *44*(3), 78–99.

ResearchGate. (2024). Technology Education. *Historical Research.*

ResearchGate. (2024). Relationship Between Student Engagement and Academic Performance. *ResearchGate.net*. Retrieved from https://www.researchgate.net/publication/381430411_Relationship_between_student_engagement_and_academic_performance

ResearchGate. (2024). An Analysis of the Research on Adaptive Learning. *ResearchGate.net.* Retrieved from https://www.researchgate.net/publication/234778584_An_analysis_of_the_research_on_adaptive_Learning_The_next_generation_of_e-learning

ResearchGate. (2024). *Adoption of Mobile Technology for Mobile Learning by University Students During COVID-19.* Retrieved from https://www.researchgate.net/publication/353271720_Adoption_of_mobile_technology_for_mobile_learning_by_university_students_during_COVID-19

ResearchGate. (2024). *Adaptive Learning: A Cluster-Based Literature Review (2011-2022).* Retrieved from https://www.researchgate.net/publication/374084207_Adaptive_learning_a_cluster-based_literature_review_2011-2022

. Sahoo, J., Mohanty, B., & Sahoo, L. (2017). Indian contribution to open access scholarly publishing in DOAJ. *Library Philosophy and Practice (e-journal), 1567.*

Selwyn, N. (2020a). The risks of digital education: From policy to practice. *Learning, Media and Technology*, *45*(2), 115–129.

Selwyn, N. (2020b). *Should robots replace teachers? AI and the future of education.* Polity Press.

Selwyn, N. (2022). The future of AI and education: Some cautionary notes. *European Journal of Education*, *57*(4), 620–631. Advance online publication. DOI: 10.1111/ejed.12532

Sergeeva, O. V., Zheltukhina, M. R., Ishmuradova, I. I., Kondakchian, N. A., Erokhova, N. S., & Zhdanov, S. P. (2025). A meta-analysis of the effectiveness of mobile-supported collaborative learning. *Online Journal of Communication and Media Technologies*, *15*(1), e202508. DOI: 10.30935/ojcmt/15948

Sharma, R., & Kumar, P. (2021). Adaptive e-learning environment based on learning styles and its impact on learners' performance. *International Journal of Educational Technology in Higher Education*, *18*(1), 1–24.

Sharma, R., & Kumar, P. (2024). *Smart Learning Environments: A Comprehensive Review on Artificial Intelligence and Learning Theories.* RSIS International Journal.

Sharma, S., Singh, G., Sharma, C. S., & Kapoor, S. (2024). Artificial intelligence in Indian higher education institutions: A quantitative study on adoption and perceptions. *International Journal of System Assurance Engineering and Management*, *15*(1), 123–135. DOI: 10.1007/s13198-023-02193-8

Siemens, G., & Long, P. (2011). Learning analytics: The future of education. *EDUCAUSE Review*, *46*(1), 30–40.

Singh, N. (2019). Perception of the ODL Learners Towards the Use of Mobile Apps in Learning in India. *Asian Journal of Distance Education, 14*(2), 71–85. https://asianjde.com/ojs/index.php/AsianJDE/article/view/425

Smiderle, R., Rigo, S. J., Marques, L. B., Coelho, J. E. P., & Jaques, P. A. (2020). The impact of gamification on students' learning, engagement, and behavior based on their personality traits. *Smart Learning Environments, 7*(3), 3. Advance online publication. DOI: 10.1186/s40561-019-0098-x

Smith, J., & Brown, L. (2022). Generative artificial intelligence and web accessibility: Towards an inclusive and sustainable future. *ResearchGate.* https://www.researchgate.net/publication/383406745_Generative_Artificial_Intelligence_and_Web_Accessibility_Towards_an_Inclusive_and_Sustainable_Future

Smith, J., Lee, P., & Kumar, R. (2024). Barriers to AI adoption in education: A faculty perspective. *Journal of Science and Educational Research, 11*(8), 211–214. https://jsaer.com/download/vol-11-iss-8-2024/JSAER2024-11-8-211-214.pdf

Solutions, P. (2024). *Challenges in Mobile Learning Adoption.* Retrieved from https://www.paradisosolutions.com/blog/top-mobile-learning-benefits/

Spector, J. M. (2019). Emerging educational technologies and research directions. *Educational Technology Research and Development, 67*(4), 789–799.

Springer (2024). *Examining the Applications of Intelligent Tutoring Systems in Real-World Educational Settings: A Meta-Analysis.* Retrieved from https://link.springer.com/article/10.1007/s10639-022-11555-x

SpringerOpen. (2024). *Feedback Mechanisms in Adaptive Learning.* Retrieved from https://slejournal.springeropen.com/articles/10.1186/s40561-024-00292-y

Sung, Y. T., Yang, J. M., & Lee, H. Y. (2017). The effects of mobile-computer-supported collaborative learning: Meta-analysis and critical synthesis. *Review of Educational Research*, *87*(4), 768–805. DOI: 10.3102/0034654317704307 PMID: 28989193

Technologies, S. M. A. R. T. (2024). SMART and the University of Melbourne partner on neurodivergent research. *SMART Technologies*. https://www.smarttech.com/about/news/smart-and-the-university-of-melbourne-partner-on-neurodivergent-research

Tenopir, C., Christian, L., & Volentine, R. (2015). Scholarly Reading by Researchers: Summary Results from Surveys in Six Countries. *Learned Publishing*, *28*(2), 124–132.

The Verge. (2024). The quickly disappearing web. *The Verge*. https://www.theverge.com/24321569/internet-decay-link-rot-web-archive-deleted-culture

TPRE. (2023). Challenges in implementing mobile learning in higher education. *Teaching and Pedagogy Research Exchange*. https://tpre.ecu.edu/index.php/tpre/article/view/756

Traxler, J. (2018). Learning with mobile devices: The future of education. *Interactive Learning Environments*, *26*(1), 4–16.

Treviranus, J. (2014). The Global Public Inclusive Infrastructure (GPII). *ResearchGate*. https://www.researchgate.net/publication/262400407_The_Global_Public_Inclusive_Infrastructure_GPII

. UNESCO. (2021). *The digital learning divide: A global perspective. UNESCO Digital Learning Report.*

UNESCO. (2021). *Education: From disruption to recovery. United Nations Educational, Scientific and Cultural Organization*. https://en.unesco.org/covid19/educationresponse

UNESCO. (2022). *Artificial intelligence in education: Challenges and opportunities for sustainable development*. United Nations Educational, Scientific and Cultural Organization.

. U.S. Department of Education. (2023). *A Retrospective on Twenty Years of Education Technology Policy.*

Van Dijk, J. (2020). *The Digital Divide*. Polity Press.

VanLehn, K. (2011). The relative effectiveness of human tutoring, intelligent tutoring systems, and other tutoring systems. *Educational Psychologist*, *46*(4), 197–221. DOI: 10.1080/00461520.2011.611369

Vasanth, S., & Sumathi, C. (2020). Learning management systems through Moodle and Google Classroom for education. *Advances in Research*, *21*(1), 32–37. DOI: 10.9734/air/2020/v21i1030249

Vij, A., & Agarwal, P. (2024). Assessing the challenges and opportunities of artificial intelligence in Indian education. *International Journal for Global Academic & Scientific Research*, *3*(1), 36–44. DOI: 10.55938/ijgasr.v3i1.71

Wang, A. I. (2015). The Wear Out Effect of a Game-Based Student Response System. *Computers & Education*, *82*, 217–227. DOI: 10.1016/j.compedu.2014.11.004

Williamson, B., & Eynon, R. (2020). Algorithmic governance and the future of education: Artificial intelligence in learning analytics. *Learning, Media and Technology*, *45*(1), 23–36. DOI: 10.1080/17439884.2020.1798995

Williamson, B., & Eynon, R. (2020). Historical perspectives on learning analytics and AI in education. *Journal of Learning Analytics*, *7*(1), 40–58.

Wobbrock, J. O., Kane, S. K., Gajos, K. Z., Harada, S., & Froehlich, J. (2011). Ability-based design: Concept, principles, and examples. *ACM Transactions on Accessible Computing*, *3*(3), 9. Advance online publication. DOI: 10.1145/1952383.1952384

. World Bank. (2023, December 2). Building inclusive education systems as a starting point—not as an afterthought. *World Bank Blogs*.

World Bank. (2023). *Digital transformation in higher education: Global trends and policy recommendations*. The World Bank Group.

. World Bank. (2023, December 2). Building inclusive education systems as a starting point—not as an afterthought. *World Bank Blogs*.

Xiao, C., Ma, W., Song, Q., Xu, S. X., Zhang, K., Wang, Y., & Fu, Q. (2024). Human-AI Collaborative Essay Scoring: A Dual-Process Framework with LLMs. *arXiv preprint arXiv:2401.06431*. Retrieved from https://arxiv.org/abs/2401.06431

YouTube. (2023). *How to Use Kahoot! for Learning Engagement.* Retrieved from https://www.youtube.com/watch?v=07_sgCii_FE

. Zawacki-Richter, O., et al. (2023). Artificial Intelligence in Intelligent Tutoring Systems Toward Sustainable Development Goals: A Review. *Smart Learning Environments.*

Zawacki-Richter, O., Marín, V. I., Bond, M., & Gouverneur, F. (2019). Systematic review of research on artificial intelligence applications in higher education—Where are the educators? *International Journal of Educational Technology in Higher Education*, *16*(1), 39. Advance online publication. DOI: 10.1186/s41239-019-0171-0

Zhang, K., & Aslan, A. B. (2021). AI technologies for education: Recent research & future directions. *Computers and Education: Artificial Intelligence*, *2*, 100025. DOI: 10.1016/j.caeai.2021.100025

Zhang, L., & Aslan, B. (2021). Blockchain in higher education: A security perspective. *International Journal of Information Security*, *15*(2), 67–85.

Zhang, P., White, J., Schmidt, D., & Lenz, G. (2018). Blockchain-based credential verification in education. *Future Internet*, *10*(3), 27.

# Chapter 2
# Artificial Intelligence and Machine Learning in Educational Apps

**Alwin T. George**
https://orcid.org/0009-0002-2824-2469
*Christ University, Bangalore, India*

**Varuna Gupta**
*Christ University, Bangalore, India*

**Vandana Mehndiratta**
*Christ University, Bangalore, India*

## ABSTRACT

*The integration of Artificial Intelligence (AI) and Machine Learning (ML) in educational applications is transforming higher education by enhancing personalized learning, intelligent tutoring, and predictive analytics. This chapter explores AI-driven functionalities, including adaptive learning, NLP, and automated assessments, while addressing challenges such as data privacy, algorithmic bias, and accessibility. Through case studies, it highlights AI's transformative potential in shaping future education, offering insights for educators, researchers, and developers interested in AI-driven learning innovations..*

DOI: 10.4018/979-8-3693-8287-5.ch002

## INTRODUCTION

The rapid advancements in technology have significantly transformed the landscape of higher education, with smart and mobile applications emerging as vital tools for facilitating learning. The swift progress of technology has notably changed higher education, especially through the incorporation of smart and mobile applications. These digital resources promote innovative learning experiences by providing tailored content delivery, enhancing student participation, and improving overall educational effectiveness (Johnson et al., 2021). Among these technological innovations, Artificial Intelligence (AI) and Machine Learning (ML) have emerged as significant forces propelling transformation in educational tools. AI and ML technologies facilitate adaptive learning, instant feedback, and intelligent tutoring, fundamentally altering how students learn and engage with educational materials (Zawacki-Richter et al., 2019).

As online education becomes more prevalent, AI-enhanced applications have played a vital role in tackling the difficulties connected with remote learning, such as engagement, accessibility, and personalized teaching (Selwyn, 2020). These applications utilize AI-powered algorithms to evaluate learning behaviors, forecast student success, and offer customized educational pathways. By utilizing technologies like Natural Language Processing (NLP), intelligent tutoring systems, and predictive analytics, AI in education significantly improves student learning experiences and aids educators in refining their teaching methods (Luckin et al., 2016).

This chapter seeks to provide a thorough examination of how AI and ML are integrated into educational applications in higher education. Specifically, it will:

- Investigate the function of AI and ML in creating smart learning environments.
- Identify essential AI-driven features such as adaptive learning, NLP, and automated evaluations.
- Examine practical uses of AI-powered educational applications and their effects on learning outcomes.

- Analyze the advantages and challenges of implementing AI in education, including ethical aspects like data privacy, algorithmic fairness, and the digital divide.
- Present case studies and empirical data demonstrating the efficacy of AI-powered educational tools.
- Offer perspectives on future trends and innovations, encompassing AI-driven virtual and augmented reality learning experiences.

By addressing these crucial topics, this chapter contributes to the ongoing dialogue regarding the significance of AI and ML in higher education. It acts as a valuable reference for educators, researchers, and developers aiming to comprehend and utilize AI-driven solutions to improve learning experiences. As AI continues to advance, its responsible and equitable application in educational contexts will be essential for shaping the future of higher education.

## Key Concepts

1. **Artificial Intelligence (AI)** – The simulation of human intelligence by machines to perform tasks like learning, reasoning, and problem-solving.
2. **Machine Learning (ML)** – A subset of AI that enables systems to learn from data and improve performance without explicit programming.
3. **Intelligent Tutoring System (ITS)** – AI-powered virtual tutors that provide real-time, customized guidance to students.
4. **Natural Language Processing (NLP)** – AI's ability to understand, interpret, and generate human language in educational applications.
5. **Predictive Analytics** – AI models that analyze student data to forecast academic performance and identify at-risk learners.
6. **Gamification in Education** – The use of AI-driven game-like elements to enhance student engagement and motivation.
7. **Blockchain in Education** – A secure digital ledger used for credential verification and academic recordkeeping.
8. **AI-Powered Career Coaching** – Machine learning models that recommend career paths and skill development courses.
9. **Ethical AI in Education** – Guidelines ensuring AI use in education is transparent, fair, and free from biases.

10. **Technology Acceptance Model (TAM)** – A framework explaining how users adopt AI-driven educational technologies based on perceived usefulness and ease of use.
11. **Digital Divide in AI Education** – The gap in access to AI-powered learning tools between privileged and underprivileged students.
12. **Data Privacy in AI Education** – Ensuring student data protection and ethical AI use in learning platforms.
13. **Blended Learning** – A mix of AI-driven instruction and traditional teaching to enhance student learning experiences.
14. **AI-Powered Microlearning** – Short, focused learning sessions powered by AI to facilitate rapid skill acquisition.
15. **AI-Enhanced Assessment Feedback** – AI-generated insights that provide students with detailed, real-time feedback on their performance.

## THEORETICAL AND CONCEPTUAL FOUNDATION

The incorporation of Artificial Intelligence (AI) and Machine Learning (ML) within educational tools is backed by various learning theories and technology adoption models that illustrate their role in improving higher education. This section examines essential frameworks that support AI-based educational applications, including Constructivist Learning Theory, Bloom's Taxonomy, Vygotsky's Zone of Proximal Development (ZPD), and the Technology Acceptance Model (TAM). These theories lay the groundwork for comprehending how AI and ML enhance learning experiences, boost engagement, and enable personalized instruction.

### 1. Constructivist Learning Theory

Constructivist Learning Theory (Piaget, 1950) posits that learners actively build knowledge through experiences and interactions rather than simply absorbing information passively. AI-infused educational applications are in harmony with this theory by providing adaptive learning experiences, wherein students interact with material customized to their progress and existing knowledge (Jonassen, 1999).

- AI-driven personalized learning environments utilize data analytics to evaluate students' comprehension levels and adjust instructions as needed (Siemens, 2013).
- AI-powered chatbots and virtual tutors (for instance, Socratic by Google) promote interaction by simulating a Socratic dialogue that fosters critical thinking (Luckin et al., 2016).

Constructivism forms the basis of adaptive learning platforms that adjust educational content in accordance with student performance, ensuring a focus on the learner.

## 2. Bloom's Taxonomy and AI-Driven Learning

Bloom's Taxonomy (Bloom et al., 1956) categorizes learning objectives into six cognitive domains: remembering, understanding, applying, analyzing, evaluating, and creating. AI-based applications support learning across these domains:

- Lower-order skills (remembering & understanding): AI-powered platforms like Quizlet and Duolingo employ ML to customize memorization and comprehension activities.
- Higher-order skills (analyzing, evaluating, creating): AI-driven writing assistants (such as Grammarly) offer feedback on critical analysis and originality, improving student writing and argumentation (Baker & Inventado, 2014).

By embedding AI within automated assessments, adaptive feedback, and intelligent tutoring, educational applications aid students in mastering advanced cognitive skills.

## 3. Vygotsky's Zone of Proximal Development (ZPD) and Intelligent Tutoring Systems

Vygotsky (1978) introduced the Zone of Proximal Development (ZPD), which indicates the interval between what a learner can accomplish alone and what they can achieve with assistance. AI-driven Intelligent Tutoring Systems (ITS) act as digital guides, providing scaffolded instruction personalized to student requirements.

- Platforms like Carnegie Learning's MATHia deliver real-time tutoring, adjusting difficulty based on student progress (Kumar et al., 2020).
- AI-driven speech recognition and NLP enhance language learning, offering feedback and support that surpasses human limitations.

By scrutinizing student behavior and performance trends, AI-powered tutoring systems guarantee targeted, data-informed instructional support that maximizes student learning potential.

## 4. Technology Acceptance Model (TAM) and AI Adoption in Education

The Technology Acceptance Model (TAM) (Davis, 1989) elucidates user acceptance of new technologies based on Perceived Usefulness (PU) and Perceived Ease of Use (PEOU). AI-driven educational applications must be developed with these aspects in mind to secure successful implementation and engagement (Venkatesh & Davis, 2000).

- Perceived Usefulness (PU): AI-driven adaptive learning platforms (e.g., Coursera and edX) boost student engagement through automated grading and AI-powered suggestions (Kumar & Sharma, 2021).
- Perceived Ease of Use (PEOU): AI chatbots and virtual assistants (for example, IBM Watson Tutor) enable smooth user interaction, alleviating cognitive load (Berkovich, 2021).

Understanding TAM aids developers in refining AI-based educational instruments for more extensive adoption in higher education.

These theoretical frameworks—Constructivist Learning Theory, Bloom's Taxonomy, ZPD, and TAM—offer vital perspectives on how

AI and ML can enrich learning, enhance engagement, and personalize education. While AI-powered educational applications hold transformative potential, their efficacy hinges on careful implementation, ensuring accessibility, addressing ethical considerations, and aligning with pedagogical objectives.

## KEY CONCEPTUAL FOUNDATIONS OF AI AND ML IN EDUCATIONAL APPLICATIONS

The combination of Artificial Intelligence (AI) and Machine Learning (ML) in educational tools is grounded in several essential conceptual bases that enhance their efficiency. These bases encompass adaptive learning, intelligent tutoring systems, natural language processing (NLP), automated evaluations, and predictive analytics. This section delves into these core concepts, supported by pertinent academic studies in educational technology.

### 1. Adaptive Learning Systems

Adaptive learning pertains to the capability of AI-driven systems to modify instructional materials dynamically according to students' learning patterns, strengths, and areas for improvement (Brusilovsky & Millán, 2007). Personalized learning platforms powered by AI evaluate student performance data in real-time and adjust resources to maximize understanding and involvement.

- Machine learning techniques monitor students' advancement and adapt the challenge level of tasks as needed, ensuring that they remain adequately stimulated without feeling overwhelmed (Johnson et al., 2021).
- Platforms such as Knewton and Carnegie Learning's MATHia leverage AI to tailor learning paths, offering personalized feedback and direction (VanLehn, 2011).

By persistently analyzing student interactions and learning habits, adaptive learning systems boost knowledge retention and improve learning outcomes (Siemens, 2013).

## 2. Intelligent Tutoring Systems (ITS)

Intelligent Tutoring Systems (ITS) function as AI-enhanced digital tutors that replicate individualized human instruction, offering tailored support and immediate feedback. These systems depend on cognitive models to evaluate student performance and provide personalized instruction strategies (Graesser et al., 2012).

- For example, the AutoTutor system utilizes Natural Language Processing (NLP) to engage students through Socratic-style conversations, encouraging critical thinking and problem-solving (Graesser, 2016).
- Case study: A case study conducted by VanLehn (2011) indicated that ITS-led learning can achieve effectiveness comparable to that of human tutoring, notably enhancing students' performance in math and science disciplines.

By pinpointing students' misunderstandings and delivering focused support, ITS promotes self-paced learning and knowledge accumulation in higher education.

## 3. Natural Language Processing (NLP) in Education

Natural Language Processing (NLP) empowers AI systems to comprehend, interpret, and produce human language, rendering it a vital asset in educational contexts (Hirschberg & Manning, 2015). Tools driven by NLP support features like automatic essay grading, chatbot tutoring, and speech recognition learning methods.

- Educational chatbots: AI-driven chatbots like Socratic by Google and IBM Watson Tutor give immediate answers to student inqui-

ries and assist in elucidating challenging concepts (Baker & Smith, 2019).
- Language learning tools: NLP allows applications like Duolingo and Babbel to evaluate pronunciation accuracy, grammar, and fluency, providing AI-generated feedback (Burstein et al., 2003).

By facilitating human-like exchanges between learners and AI systems, NLP boosts engagement and learning effectiveness in online education settings (Huang et al., 2020).

## 4. Automated Assessments and AI-Powered Feedback

Automated assessment tools driven by AI simplify grading, evaluation, and feedback processes, alleviating the burden on educators while guaranteeing uniformity and impartiality (Balfour, 2013). These tools utilize ML algorithms to analyze written replies, multiple-choice items, and even subjective essays.

- Automated Essay Scoring (AES): Platforms like Turnitin and Grammarly employ AI to assess writing for grammar, coherence, and originality, providing prompt feedback (Shermis & Burstein, 2013).
- AI-driven formative evaluations: ML models scrutinize student responses to identify learning deficiencies and recommend specific remedial actions (Shute & Rahimi, 2021).

Research shows that AI-assisted grading diminishes biases prevalent in human assessments while enhancing grading efficiency and expediting student feedback response times (Luo et al., 2019).

## 5. Predictive Analytics for Student Performance

In education, predictive analytics utilizes AI and ML models to anticipate student performance, identify learners at risk, and facilitate early interventions (Siemens & Long, 2011).

- Early warning systems: AI examines historical student data, attendance patterns, and engagement metrics to pinpoint students who might have academic difficulties (Bienkowski et al., 2012).
- Personalized learning pathways: Platforms such as Coursera and edX utilize AI-based recommendation systems to propose courses and study resources tailored to learner preferences and previous interactions (Pardo et al., 2016).

Research indicates that AI-enhanced predictive analytics can boost student retention rates and enable institutions to proactively assist at-risk learners, thus enhancing overall academic achievement (Rienties et al., 2016).

The theoretical underpinnings of AI and ML in educational contexts—adaptive learning, intelligent tutoring systems, NLP, automated assessments, and predictive analytics—revolutionize higher education by improving engagement, personalization, and efficiency. Yet, these advancements must be implemented thoughtfully to tackle ethical issues, data privacy, and accessibility challenges. By incorporating AI-based educational tools within these conceptual frameworks, institutions can enhance learning experiences and equip students for a digital future.

## KEY CONCEPTUAL FOUNDATIONS OF AI AND ML IN EDUCATIONAL APPLICATIONS

Artificial Intelligence (AI) and Machine Learning (ML) are revolutionizing the educational landscape by improving adaptive learning, intelligent tutoring, natural language processing (NLP), automated assessments, and predictive analytics. These technologies customize instruction, enhance student engagement, and maximize learning results. Nevertheless, their success hinges on the theoretical frameworks and conceptual models that inform their creation and use. This section examines essential conceptual foundations, offering a critical analysis of existing frameworks and their shortcomings.

## 1. Adaptive Learning and Constructivist Theory

### • Conceptual Foundation

Adaptive learning platforms utilize AI and ML to customize educational content according to students' performance, learning preferences, and progression (Brusilovsky & Millán, 2007). These systems are in harmony with Constructivist Learning Theory, which asserts that knowledge is built through active participation rather than through passive absorption (Piaget, 1950; Jonassen, 1999).

AI-enabled adaptive systems employ real-time analytics to adjust learning pathways in real-time, ensuring that students receive instruction appropriate to their cognitive levels (Siemens, 2013). For instance:

Knewton and Carnegie Learning's MATHia evaluate student responses to modify content difficulty (VanLehn, 2011).

AI-based recommendation systems propose learning resources based on prior interactions (Pardo et al., 2016).

### • Critical Analysis

While adaptive learning promotes personalized education, it falls short in providing the social and emotional intelligence that human instructors offer (Selwyn, 2020). Moreover, worries about data privacy and algorithmic biases persist, as AI models might reinforce disparities in educational access (Zawacki-Richter et al., 2019).

## 2. Intelligent Tutoring Systems (ITS) and Zone of Proximal Development (ZPD)

### • Conceptual Foundation

Intelligent Tutoring Systems (ITS) serve as AI-powered virtual tutors that deliver scaffolded instruction, immediate feedback, and tailored learning support (Graesser et al., 2012). This model corresponds with Vygotsky's

Zone of Proximal Development (ZPD), which posits that learners can reach deeper understanding with appropriate guidance (Vygotsky, 1978).

Examples include:

AutoTutor employs NLP to replicate human tutoring by engaging students in Socratic dialogue (Graesser, 2016).

Socratic by Google offers AI-generated step-by-step assistance for problem-solving.

• Critical Analysis

Although ITS simulates human tutoring, it lacks the emotional intelligence and motivational techniques inherent to human educators (Baker & Smith, 2019). Furthermore, excessive dependence on AI tutors may hinder students' capability to develop self-regulated learning skills (VanLehn, 2011).

## 3. Natural Language Processing (NLP) and Cognitive Load Theory

• Conceptual Foundation

NLP facilitates AI-powered chatbots, virtual assistants, and automated feedback systems that enable smooth interactions for students (Hirschberg & Manning, 2015). Sweller (1988) illustrates how educational tools based on NLP can alleviate cognitive overload by streamlining the delivery of complex information.

Duolingo and Babbel apply NLP for speech recognition and provide real-time support for language learning (Burstein et al., 2003).

AI-assisted essay grading systems, such as Turnitin and Grammarly, offer automated assessments of writing structure, grammar, and coherence (Shermis & Burstein, 2013).

- Critical Analysis

Despite the effectiveness of NLP, AI-based systems frequently encounter challenges with contextual comprehension and subjectivity, especially in grading creative writing and assessing the quality of arguments (Huang et al., 2020). Additionally, biases present in NLP models can disadvantage non-native speakers, resulting in inequitable evaluations (Baker & Inventado, 2014).

## 4. Automated Assessments and AI-Powered Feedback

- Conceptual Foundation

AI-enhanced assessments utilize Machine Learning algorithms to assess student performance and provide feedback instantaneously (Balfour, 2013). This aligns with Formative Assessment Theory, which stresses the importance of ongoing feedback to enhance learning (Black & Wiliam, 1998).

Automated Essay Scoring (AES) tools evaluate syntax, coherence, and the structure of arguments (Shermis & Burstein, 2013).
AI-based multiple-choice assessments enhance grading efficiency and lessen the workload for educators (Luo et al., 2019).

- Critical Analysis

AI performs well in grading objective assessments but faces challenges with subjective tasks and higher-order thinking (Shute & Rahimi, 2021). Furthermore, there are concerns regarding fairness in algorithms and potential biases in grading, especially with AI-scored essays (Berkovich, 2021).

## 5. Frameworks for Predictive Analytics and Learning Analytics

### • Conceptual Foundation

In educational contexts, predictive analytics utilizes AI and ML to anticipate student outcomes, pinpoint learners at risk, and propose interventions (Bienkowski et al., 2012). This approach is based on Learning Analytics Frameworks, which prioritize decision-making driven by data (Siemens & Long, 2011).

Early warning systems can identify academic challenges and recommend intervention strategies (Rienties et al., 2016).

Analytics related to student engagement, powered by AI, assist institutions in refining curriculum design (Pardo et al., 2016).

### • Critical Discussion

Although predictive analytics has the potential to enhance student retention, its success hinges on the accuracy of data inputs. Biases present in training datasets can result in erroneous predictions, which may negatively impact certain demographic groups of students (Zawacki-Richter et al., 2019).

The underlying concepts of AI and ML in educational settings such as adaptive learning, intelligent tutoring, natural language processing, automated assessments, and predictive analytics—contribute to improved student engagement, personalization, and overall efficiency. Nevertheless, each framework also has limitations, including ethical issues, biases, and risks of excessive dependence on AI. In order to fully leverage the advantages of AI in education, it is essential for stakeholders to prioritize responsible implementation, ensure equity, and continuously improve AI models.

## CONCEPTUAL DEVELOPMENT:

The incorporation of Artificial Intelligence (AI) and Machine Learning (ML) in educational tools is based on established learning theories and new technological frameworks. This section articulates the view that AI-driven educational resources improve personalized learning, increase student engagement, and enhance assessment accuracy, while also recognizing ethical, pedagogical, and implementation issues. The premise is that when designed and implemented effectively, AI-powered educational applications can notably boost educational outcomes in higher education.

### 1. AI as a Catalyst for Personalized Learning: A Constructivist Perspective

- Argument Development

Constructivist Learning Theory (Piaget, 1950; Vygotsky, 1978) highlights that learning takes place when students interact with content and build knowledge actively. AI conforms to this perspective by creating adaptive learning settings that adjust to individual educational needs.

Machine learning algorithms monitor student progress and adjust content in real time (Brusilovsky & Millán, 2007).

AI-driven suggestions tailor coursework, ensuring that learners receive materials appropriate for their proficiency levels (Siemens, 2013).

- Hypothesis

H1: AI-powered adaptive learning platforms boost student engagement and knowledge retention by offering personalized educational experiences.

- Counterpoint

Although AI customizes learning, it may lack the social and emotional intelligence of human educators, which could diminish opportunities for collaborative and social learning (Selwyn, 2020). Tackling this limitation necessitates blended learning approaches, where AI augments human teaching rather than substitutes it.

- Logical Explanation

Constructivist Learning Theory (Piaget, 1950; Vygotsky, 1978) posits that the most effective learning occurs when students are actively involved with the material, building knowledge through their experiences. AI-powered adaptive learning technologies exemplify this theory by delivering personalized educational experiences that modify in real time according to student performance and learning behaviors.

- Reasoning

Conventional education adheres to a uniform approach, providing the same instructional materials to all students, regardless of their previous knowledge, learning pace, or understanding level. AI addresses this issue by:

Monitoring student advancement and adjusting content on-the-fly.
Offering tailored recommendations that cater to individual learning preferences.
Delivering instant feedback to reinforce concepts prior to moving forward.

- Examples

Knewton and Carnegie Learning's MATHia: These platforms utilize machine learning to evaluate student answers and modify the difficulty of content accordingly, ensuring that learners encounter an appropriate level of challenge (VanLehn, 2011).

Duolingo's AI-driven language learning: Employs natural language processing and machine learning to customize exercises, enabling learners to concentrate on their weaker areas while solidifying their strengths (Burstein et al., 2003).

## 2. AI-Powered Intelligent Tutoring Systems (ITS) and Cognitive Scaffolding

### • Argument Development

According to Vygotsky's Zone of Proximal Development (ZPD), students learn most effectively when provided with guided support that helps them advance from their existing level of understanding to higher levels (Vygotsky, 1978). AI-based Intelligent Tutoring Systems (ITS) serve as virtual mentors, delivering real-time feedback and step-by-step assistance.

Example: AutoTutor emulates human tutoring through Natural Language Processing (NLP) to facilitate Socratic-style learning (Graesser et al., 2012).

Case Study: VanLehn (2011) discovered that learning through ITS significantly enhances student performance, especially in math and language studies.

### • Hypothesis

H2: AI-driven Intelligent Tutoring Systems improve students' comprehension and problem-solving abilities by offering personalized, real-time guidance.

### • Counterpoint

ITS lacks the emotional intelligence and adaptability that human tutors provide in terms of student motivation. Excessive dependence on AI tutors might also hinder the development of self-regulated learning skills, leading students to rely on automated responses instead of fostering independent problem-solving capabilities (Baker & Smith, 2019).

• Logical Explanation

Vygotsky's Zone of Proximal Development (ZPD) suggests that learners enhance their performance with supported guidance that connects their existing skills to their potential understanding (Vygotsky, 1978). AI-driven Intelligent Tutoring Systems (ITS) serve as virtual guides, assisting students progressively with immediate explanations, hints, and evaluations.

• Reasoning

Conventional classrooms frequently find it challenging to give each student personal attention due to large student-teacher ratios. ITS addresses this issue by:

Evaluating student responses and anticipating learning obstacles.
Delivering prompt, interactive feedback akin to that of a human tutor.
Offering tailored problem-solving techniques based on student outcomes.

• Examples

AutoTutor: Employs natural language processing to replicate human dialogue and Socratic questioning, aiding students in enhancing their critical thinking skills (Graesser et al., 2012).
Socratic by Google: An AI-driven application that examines student inquiries and delivers step-by-step explanations for problem-solving.

## 3. AI-Driven Assessments and the Challenge of Fairness

• Argument Development

AI-powered automated assessment systems enhance grading and feedback efficiency, thereby mitigating human bias (Balfour, 2013). This coincides with Formative Assessment Theory, which stresses the importance of ongoing feedback to support student development (Black & Wiliam, 1998).

Automated Essay Scoring (AES) tools like Turnitin and Grammarly offer immediate feedback on grammar, coherence, and originality (Shermis & Burstein, 2013).

AI-powered multiple-choice assessments facilitate large-scale evaluations with prompt feedback (Luo et al., 2019).

- Hypothesis

H3: AI-powered automated assessment systems enhance grading efficiency and fairness by minimizing human biases and providing instant feedback.

- Counterpoint

Despite its benefits, AI encounters difficulties in assessing creativity, critical thinking, and nuanced arguments (Shute & Rahimi, 2021). Additionally, AI inherently mirrors biases present in its training data, which can disadvantage non-native speakers or marginalized groups (Baker & Inventado, 2014). Addressing these challenges necessitates ongoing algorithmic improvements and human oversight.

- Logical Explanation

Automated assessment tools enhance grading efficiency by removing human bias and providing uniformity. These tools are in accordance with Formative Assessment Theory, which focuses on continuous feedback to boost learning outcomes (Black & Wiliam, 1998).

- Reasoning

Conventional assessments are typically:

Labor-intensive: Evaluating essays and open-ended responses can slow down the feedback process.

Subject to human bias: Grading subjectivity can create inconsistencies.

AI-driven assessments address these challenges by:

Utilizing NLP to evaluate essay organization, clarity, and grammatical correctness.
Using ML models to detect learning deficiencies and suggest targeted solutions.

• Examples

Turnitin and Grammarly: AI-powered writing tools that assess clarity, originality, and grammar, offering students prompt feedback (Shermis & Burstein, 2013).
AI-assisted multiple-choice evaluations: Platforms such as Coursera and edX leverage AI to automatically score quizzes and assignments, ensuring rapid and equitable assessment.

## 4. Predictive Analytics and Early Intervention for At-Risk Students

• Argument Development

AI-powered predictive analytics can evaluate patterns of student engagement, performance metrics, and behavioral data to detect at-risk students at an early stage (Siemens & Long, 2011). This is consistent with Learning Analytics Frameworks, which prioritize data-informed decision-making to promote student success (Bienkowski et al., 2012).

Example: Institutions that implement AI-driven early warning systems have seen improvements in student retention rates (Rienties et al., 2016).
AI-based academic interventions offer tailored study suggestions based on student progress (Pardo et al., 2016).

- Hypothesis

H4: AI-powered predictive analytics enhance student retention and academic achievement by facilitating timely support for at-risk learners.

- Counterpoint

Predictive analytics depend on historical data, which might perpetuate systemic biases if AI models aren't appropriately calibrated (Zawacki-Richter et al., 2019). Additionally, ethical issues regarding data privacy and monitoring in education must be addressed to avoid the misuse of student information.

- Logical Explanation

Predictive analytics powered by AI examine student engagement habits, previous performance, and behavioral trends to detect students at risk of failing before it happens. This is in harmony with Learning Analytics Frameworks, which support the use of data-driven interventions (Siemens & Long, 2011).

- Reasoning

Many educational institutions face challenges with high dropout rates due to insufficient early intervention approaches. AI provides a solution by:

Monitoring student engagement and progress in real-time.
Recognizing behavioral patterns associated with academic risk (such as high absenteeism and decreasing test scores).
Proposing individualized interventions like tutoring or modified coursework.

- Examples

Early Warning Systems: Colleges employing AI-enabled risk identification have seen improvements in retention rates (Rienties et al., 2016).

AI-assisted academic interventions: Systems such as Purdue's Course Signals assess student behavior and recommend specific support strategies.

This conceptual overview highlights both the potential advantages and challenges associated with AI and ML in educational contexts. The proposed hypotheses indicate that AI-driven learning platforms can improve engagement, understanding, assessment efficiency, and student retention. Nonetheless, issues pertaining to bias, fairness, emotional intelligence, and ethical considerations underscore the necessity for responsible AI application. Future studies should investigate hybrid AI-human models to ensure that AI enhances traditional teaching methods instead of entirely replacing them.

## POTENTIAL IMPLICATIONS AND APPLICATIONS OF THE CONCEPTUAL MODEL

The conceptual framework surrounding AI and ML-based educational tools carries profound implications for higher education, policy formulation, workforce readiness, and ethical governance of AI technologies. By adopting AI-driven learning solutions, educational institutions can enhance customized learning experiences, boost student involvement, streamline evaluation processes, and forecast academic outcomes. Nevertheless, the reach of these technologies goes beyond conventional classrooms, affecting decision-making within institutions, digital transformation initiatives, and the broader educational landscape.

### 1. Implications for Higher Education Institutions

- Customized Learning at Scale

Implication: AI-enhanced adaptive learning frameworks transform conventional education by customizing content delivery according to individual student requirements. This transitions learning from a generalized approach to one that is centered on the needs of each student.

Application:

- Smart Learning Platforms: Schools can implement systems such as Knewton, Coursera, and edX, which modify course materials in real-time in response to student progress (Siemens, 2013).
- Dynamic Curriculum Design: AI-generated insights enable educators to revise curricula based on student performance patterns, ensuring a syllabus that adapts progressively to learner requirements.

Challenges and Considerations:

- Ensuring fair access to AI-based resources to avoid exacerbating the digital divide (Zawacki-Richter et al., 2019).
- Balancing AI-driven personalization with conventional teaching methods to maintain human interaction.

## • AI-Driven Institutional Decision-Making

Implication: AI can aid universities and educational policymakers in making informed decisions regarding student success, faculty effectiveness, and resource distribution.

Application:

- Enrollment Management: Predictive analytics can reveal patterns in student applications, retention risks, and academic achievement, helping institutions refine their recruitment and retention efforts (Rienties et al., 2016).
- AI for Faculty Support: AI-generated insights assist universities in evaluating teaching effectiveness, student input, and faculty workloads, facilitating performance evaluations and faculty development initiatives.

Challenges and Considerations:

- Ethical issues related to data privacy and surveillance necessitate universities to formulate clear data governance policies.
- The potential for algorithmic bias in decision-making calls for ongoing monitoring and validation of AI systems.

## 2. Implications for Student Learning and Assessment

### ▪ Automated and Equitable Assessment Systems

Implication: AI-assisted grading and feedback tools offer immediate assessments, lessen instructor burden, and reduce grading biases.

Application:

- Automated Essay Scoring (AES): Tools like Turnitin, Grammarly, and WriteLab evaluate students' writing for grammatical accuracy, coherence, and originality (Shermis & Burstein, 2013).
- Instant Formative Feedback: AI-enabled multiple-choice tests and natural language processing (NLP) chatbots provide instant grading and individualized feedback.

Challenges and Considerations:

- AI encounters difficulties assessing subjective areas such as creativity, argumentative skills, and ethical judgment (Shute & Rahimi, 2021).
- The danger of perpetuating biases in AI grading systems emphasizes the need for human oversight and fairness assessments.

### • AI-Powered Learning Assistants

Implication: AI chatbots and Intelligent Tutoring Systems (ITS) can function as 24/7 educational aides, enhancing accessibility and promoting self-directed learning.

Application:

- AI Tutors: Platforms such as AutoTutor and Google's Socratic engage students through Socratic questioning and problem-solving techniques (Graesser et al., 2012).
- AI-Powered Chatbots: Educational institutions can use AI-driven chatbots to address academic questions, offer study materials, and support mental health (Baker & Smith, 2019).

Challenges and Considerations:

- Limited emotional intelligence: AI tutors are unable to completely replicate the mentorship and emotional backing provided by humans.
- Student dependency on AI may hinder critical thinking skills and self-regulated learning (VanLehn, 2011).

## 3. Implications for Workforce Preparation and Lifelong Learning

### • Closing the Gap Between Education and Industry

Implication: Education powered by AI aligns academic training with the needs of the industry, equipping students for workplaces that utilize AI technologies.

Application:

- Skill-Based Learning Analytics: AI monitors the progress of students' skill acquisition and suggests career-focused learning paths (Siemens & Long, 2011).
- AI-Powered Career Guidance: Tools such as LinkedIn Learning and Coursera Career Coach leverage machine learning to connect students with job opportunities and personalize career development strategies.

Challenges and Considerations:

- The ever-evolving job market necessitates that AI models be regularly updated.
- Ethical issues related to AI in hiring processes may lead to biased algorithms affecting the equity of recruitment practices.

### • Lifelong Learning and Reskilling Initiatives

Implication: AI promotes ongoing education for professionals by providing on-demand skill development opportunities.

Application:

- Corporate AI Training: Organizations utilize AI-driven platforms like Udacity, IBM SkillsBuild, and Google AI Academy to reskill their workforce in areas such as data science, automation, and AI ethics.

- AI-Driven Microlearning: Brief, AI-created learning modules assist professionals in quickly acquiring new skills in rapidly evolving sectors.

Challenges and Considerations:

- It is crucial to ensure that AI-based learning remains both accessible and affordable.
- There is a risk of AI monopolization in education, where large corporations could control educational pathways.

## 4. Implications for Ethical AI Governance in Education

### • Prioritizing AI Ethics and Data Privacy

Implication: With AI gathering extensive student data, it is essential for institutions to focus on ethical AI practices and effective data protection measures.

Application:

- Regulatory Compliance: Educational institutions must adhere to GDPR, FERPA, and other relevant laws to protect student privacy.
- Bias Audits for AI Models: Schools and universities should routinely perform fairness and bias evaluations to ensure AI does not reinforce discrimination (Zawacki-Richter et al., 2019).

Challenges and Considerations:

- The decision-making process driven by AI needs to be transparent, preventing opaque algorithms from making unmonitored decisions about students' academic careers.
- Institutions are required to inform students about how AI algorithms operate and what data is being collected.

The integration of AI in education relies on responsible advancement, appropriate regulatory measures, and a dedication to creating inclusive learning settings. AI should enhance the experiences of both learners and educators instead of dictating those experiences. Through careful application, AI-driven education has the potential to democratize access to knowledge and contribute to a more just, data-informed educational landscape.

## IMPLICATIONS AND FUTURE RESEARCH: THE IMPACT OF AI AND ML IN EDUCATIONAL APPLICATIONS

The combination of Artificial Intelligence (AI) and Machine Learning (ML) in educational applications offers transformative possibilities for higher education, instructional design, assessment strategies, and institutional decision-making. However, these advancements also bring about challenges, ethical dilemmas, and research gaps that need further inquiry. This section discusses the potential effects of AI-enhanced education and suggests future research avenues to promote responsible, effective, and equitable use of AI in higher education.

### 1. Implications for the Field and Practice

#### 1.1. Advancing Personalized Learning and Engagement

- Potential Impact: AI-driven adaptive learning systems could significantly transform education by providing customized learning experiences suited to individual students' needs, learning styles, and progress (Brusilovsky & Millán, 2007).
- Key Contributions:

  Increased Student Engagement: AI-enhanced gamification, intelligent tutoring, and instantaneous feedback can boost motivation and participation (Siemens, 2013).

  Improved Learning Outcomes: AI-enabled recommendation systems can propose personalized resources, helping students effectively understand complex subjects (Pardo et al., 2016).

  Accessibility for Diverse Learners: AI technologies such as speech recognition, translation tools, and assistive technologies can improve access for students with disabilities.

  - Challenges and Considerations:

Risk of becoming too dependent on AI, which may diminish the role of human mentorship and collaborative learning.

The digital divide may hinder students from low-income backgrounds from accessing AI-driven learning tools (Zawacki-Richter et al., 2019).

## 1.2. AI-Enhanced Assessments and Learning Analytics

- Potential Impact: AI-based assessments can improve grading efficiency, minimize human biases, and deliver immediate, data-driven feedback.
- Key Contributions:

  Automated Essay Scoring (AES): AI systems can assess student writing in real time, enhancing grading consistency (Shermis & Burstein, 2013).

  Predictive Analytics for Academic Success: AI-powered learning analytics can aid in identifying at-risk students early on and provide tailored intervention strategies (Bienkowski et al., 2012).

  Scalable, Cost-Effective Assessment: AI can alleviate administrative tasks for educators, allowing them to focus on curriculum development and student mentorship.
- Challenges and Considerations:

  Equity in AI grading: AI may struggle to evaluate creativity, argumentation, and ethical reasoning (Shute & Rahimi, 2021).

  Bias in Predictive Models: AI systems trained on biased datasets may disproportionately identify certain student demographics as at-risk (Baker & Inventado, 2014).

## 1.3. Institutional Decision-Making and Policy Development

- Potential Impact: AI-driven insights can assist universities and policymakers in optimizing resource distribution, curriculum development, and student support services.
- Key Contributions:

AI for Institutional Planning: Predictive analytics can guide enrollment management, faculty performance evaluations, and strategic investments (Rienties et al., 2016).

Data-Driven Policy Recommendations: AI models can evaluate student feedback, course performance, and faculty workload, aiding policymakers in making informed decisions.

- Challenges and Considerations:

  Privacy and Ethical Issues: Institutions need to comply with data protection laws (e.g., GDPR, FERPA) to protect student data (Zawacki-Richter et al., 2019).

  Transparency in AI Decision-Making: Decisions made by AI in institutions must be interpretable and auditable to prevent unaccountable black-box algorithms.

## 2. Future Research Directions

### 2.1. Tackling AI Bias and Ethical Issues

AI models can carry biases from their training sets, resulting in unjust outcomes in grading, predictive analytics, and learning recommendations (Zawacki-Richter et al., 2019).

- Key Research Questions:

  How can AI models be developed using diverse, unbiased datasets to ensure fairness?

  What regulatory frameworks should be established to guide AI-driven decision-making in education?

  How can assessments powered by AI be structured to fairly evaluate creativity and critical thinking?

### 2.2. Investigating Hybrid AI-Human Learning Models

Even with AI's advantages, human educators play a crucial role in education. Research should examine blended learning models where AI complements rather than replaces human teachers (Selwyn, 2020).

- Key Research Questions:
  What should be the ideal balance between AI automation and human instruction?
  How can AI be developed to support, instead of substituting, human mentorship?
  What best practices should be followed for integrating AI-driven tutors into conventional classrooms?

## 2.3. AI for Lifelong Learning and Workforce Development

With the rapid evolution of industries, it is essential for AI-enhanced education to align with job market requirements, ensuring that students acquire the skills pertinent to future employment.

- Key Research Questions:
  In what ways can AI-driven learning platforms assist professionals in acquiring new skills and adapting throughout their careers?
  What functions should AI serve in tailoring career development paths?
  How can universities collaborate with industry to jointly create AI-focused educational programs?

## 2.4. Ensuring AI Accessibility and Bridging the Digital Divide

Education powered by AI should be made available to all learners, irrespective of their socioeconomic status. Upcoming research needs to tackle disparities in the availability of AI-powered resources (Siemens & Long, 2011).

- Key Research Questions:
  In what manner can AI technologies be designed and introduced at low costs for disadvantaged students?
  What regulations can be implemented to guarantee fair access to AI-powered educational tools across various regions?
  What responsibilities should governments and private sectors have in funding educational initiatives that utilize AI?

Integrating AI and ML into education has significant implications, providing tailored learning experiences, automated evaluations, informed decision-making, and alignment with workforce needs. Nevertheless, forthcoming research must focus on the ethical use of AI, the presence of bias, accessibility issues, and finding the right balance between automation and human teaching to fully leverage AI's advantages while promoting equitable and inclusive educational experiences.

## AREAS FOR FURTHER RESEARCH AND EXPLORATION

The theoretical framework of AI and ML in educational settings underscores their transformative capabilities in tailoring learning experiences, providing intelligent tutoring, automating evaluations, and employing predictive analytics. Nonetheless, several gaps persist, necessitating deeper investigation to ensure that AI is implemented ethically, effectively, and inclusively. The following are critical areas for upcoming research:

### 1. Addressing AI Bias and Fairness in Education

• Significance

AI-driven educational tools develop insights from existing datasets, which may harbor biases related to race, gender, socioeconomic factors, or learning capacities. If these biases are left uncorrected, they can reinforce disparities in grading practices, student assistance, and educational opportunities (Zawacki-Richter et al., 2019).

• Research Questions

How can AI systems be developed to identify and reduce biases in educational evaluations?

What measures can be enacted to guarantee equitable AI-based student support?

How do various demographic groups perceive AI-enhanced learning differently?

• Possible Methodologies

Conducting bias evaluations of AI assessment frameworks to analyze patterns of inequitable results.

Comparative research examining AI-facilitated versus human-led evaluations across a variety of student demographics.

Creating AI fairness protocols to establish ethical guidelines for utilizing AI in educational contexts.

## 2. AI's Contribution to Fostering Critical Thinking and Creativity

• Significance

While AI is skilled in adaptive learning and personalized feedback, its competency in promoting higher-order cognitive abilities such as critical thinking, creativity, and ethical reasoning is still inadequately explored (Siemens, 2013).

• Research Questions

Do AI-enhanced learning tools promote deep understanding rather than mere information memorization?

Which AI-driven techniques can further problem-solving and creative thought processes?

What impact does AI have on student motivation and inquisitiveness?

• Possible Methodologies

Experimental research contrasting AI-supported learning with conventional teaching methodologies for creative disciplines.

Cognitive evaluations of student involvement with AI-provided feedback.

Case studies on AI-assisted brainstorming applications in education.

## 3. The Collaboration between AI and Human Educators: Striking a Balance

### • Significance

AI can support teachers but cannot substitute for human guidance, emotional awareness, and interpersonal learning opportunities (Selwyn, 2020). Achieving the proper mix of AI efficiency and human teaching is critical.

### • Research Questions

What is the optimal distribution of responsibilities between AI tools and human instructors in educational environments?

How does the support of AI in teaching influence teachers' workload and the quality of instruction?

In what ways can AI be utilized to train educators and enhance teaching practices?

### • Possible Methodologies

Surveys and interviews with teachers regarding AI's function in lesson design and evaluation.

Observational research comparing student achievement in AI-supported versus traditional teaching settings.

Evaluations of AI-based programs for teacher training and pedagogical improvement.

## 4. Utilizing AI for Predictive Analytics and Early Alerts

### • Significance

Predictive analytics has the potential to detect students at risk early on, facilitating timely interventions. However, issues related to data accuracy, false positives, and ethical concerns about student profiling remain a challenge (Bienkowski et al., 2012).

- Research Questions

How can predictive analytics effectively gauge student risk without reinforcing existing biases?
What are the ethical considerations surrounding AI-driven student profiling?
How successful are AI-enhanced interventions in boosting student retention and achievement?

- Possible Methodologies

Longitudinal studies monitoring student outcomes following AI-led interventions.
Research on algorithm transparency to ensure equitable and unbiased decision-making.
A comparative study of assessments of academic risk determined by human judgment versus AI predictions.

## 5. AI-Based Assessment and Alternative Evaluation Approaches

- Significance

Although AI is capable of grading multiple-choice tests and structured essays, its proficiency in assessing critical thinking, originality, and ethical reasoning is still in doubt (Shute & Rahimi, 2021).

- Research Questions

How can AI-facilitated assessments reliably evaluate student creativity?
Is it possible to design AI to offer nuanced, context-sensitive feedback in more subjective areas?
How do learners feel about assessments graded by AI compared to those graded by humans?

• Possible Methodologies

Qualitative studies exploring student experiences with AI-based grading tools.

Comparative analyses of grading outcomes between human and AI evaluations focusing on consistency and fairness.

Improvements in algorithms related to natural language processing for more sophisticated essay assessment.

## CASE STUDIES AND REAL-WORLD APPLICATIONS OF AI AND ML IN EDUCATIONAL APPS

### 1. Duolingo: AI-Enhanced Personalized Language Learning

Duolingo, a widely recognized platform for language acquisition, utilizes AI and ML to create tailored learning paths for students. Its AI-driven chatbots, adaptive learning algorithms, and NLP-focused speech recognition allow for real-time customization of the learning experience (Settles & Meeder, 2016).

• Key AI Features:

Tailored Content Adjustment: The ML algorithms modify question difficulty according to learner performance.

Speech Analysis & Feedback: NLP aids students in refining their pronunciation by assessing speech patterns.

Gamified Engagement: AI-driven progress tracking boosts learner motivation.

• Impact on Education:

Research indicates that spending 34 hours on Duolingo equates to a full university semester of language studies (Vesselinov & Grego, 2012).

Duolingo's adaptive AI model enhances learning effectiveness, making language study more accessible and scalable.

## 2. Carnegie Learning's MATHia: AI-Driven Customized Math Tutoring

Carnegie Learning's MATHia is an AI-powered Intelligent Tutoring System (ITS) aimed at providing personalized math instruction for students in higher education (VanLehn, 2011).

- Key AI Features:

Continuous Student Progress Monitoring: AI algorithms identify learning gaps and adjust educational content in real-time.
Step-by-Step Problem Assistance: The system offers instant feedback, replicating guidance from a human instructor.
Data-Based Insights for Educators: Teachers gain access to real-time analytics concerning student progress.

- Impact on Education:

Studies revealed that students utilizing MATHia performed notably better on standardized math assessments than peers in conventional classroom settings (Koedinger et al., 2013).
The AI system alleviates math-related anxiety by providing a personalized, student-focused tutoring approach.

## 3. Coursera and edX: AI-Enabled Learning Analytics

Massive Open Online Courses (MOOCs) such as Coursera and edX employ AI to tailor course recommendations, automate grading processes, and deliver real-time learning analytics (Pardo et al., 2016).

• Key AI Features:

Customized Course Recommendations: AI suggests courses based on learner interests and previous engagements.

Automated Evaluation: AI-driven Natural Language Processing (NLP) technologies assess essays and quizzes.

Predictive Analytics for Learner Success: AI recognizes students at risk and proposes early interventions.

• Impact on Education:

Coursera's AI platform examines learning trends to enhance course material and student interaction (Siemens & Long, 2011).

edX experienced a 20% increase in student retention through AI-enabled early intervention strategies (Reich, 2020).

## 4. IBM Watson Tutor: AI-Enhanced Cognitive Tutoring

IBM Watson's AI-based virtual tutor is employed in higher education institutions to improve student learning experiences (Berkovich, 2021).

• Key AI Features:

Interactive Conversational AI: Watson Tutor utilizes NLP to engage with students, resolving inquiries and facilitating discussions.

Adaptive Questioning Techniques: AI modifies teaching approaches based on student interactions.

Immediate Academic Assistance: AI chatbots offer round-the-clock learning support.

• Impact on Education:

The utilization of IBM Watson Tutor in STEM disciplines has led to heightened student engagement and quicker learning retention (Luckin et al., 2016).

Universities implementing IBM Watson have reported a 30% decrease in faculty workload, enabling professors to concentrate on more complex teaching tasks.

## 5. Purdue University's Course Signals: AI-Based Student Retention Initiative

Purdue University created Course Signals, an AI-powered predictive analytics tool designed to detect students at risk and suggest interventions (Arnold & Pistilli, 2012).

• Key AI Features:

Proactive Warning System: AI forecasts potential student dropouts by analyzing behavioral data.
Tailored Support: Educators receive immediate alerts to facilitate early intervention.
Insights into Student Performance: AI scrutinizes attendance, engagement, and coursework data.

• Impact on Education:

A study from Purdue found that students receiving AI-driven interventions achieved a 15% increase in course completion rates.
Course Signals demonstrated that early intervention strategies enhance academic performance and retention statistics (Rienties et al., 2016).

## KEY INSIGHTS AND CONTRIBUTIONS

This chapter has examined the transformative impact of Artificial Intelligence (AI) and Machine Learning (ML) within educational settings, with an emphasis on personalized learning, intelligent tutoring, automated assessments, predictive analytics, and decision-making at the institutional level. The essential insights and contributions from this chapter are outlined below:

## 1. Theoretical and Conceptual Foundations

- AI-enhanced education is in harmony with Constructivist Learning Theory (Piaget, 1950), which stresses the importance of active, learner-centered experiences.
- Vygotsky's Zone of Proximal Development (ZPD) is foundational for AI-supported Intelligent Tutoring Systems (ITS), which offer scaffolded, real-time help to learners.
- Bloom's Taxonomy (1956) provides a framework for creating AI instruments that strengthen both lower-order (such as memorization) and higher-order (like analysis and creation) cognitive abilities.
- The Technology Acceptance Model (TAM) elucidates how students and educators embrace AI-driven learning aids, emphasizing key factors like perceived usefulness and ease of use.

Key Contribution: AI technologies are in alignment with established educational theories, enhancing personalized and data-driven learning.

## 2. AI-Powered Personalized Learning and Student Engagement

- AI-based adaptive learning systems modify content in response to individual student progress, which enhances knowledge retention and engagement (Hardan & Ghabayen, 2025).
- For example: Knewton and Carnegie Learning's MATHia dynamically adapt content, ensuring students receive focused instruction.
- AI-driven gamification and immediate feedback boost motivation, especially in self-directed learning contexts.

Key Contribution: AI allows for personalized education on a large scale, promoting greater engagement and effectiveness compared to conventional approaches.

## 3. AI-Driven Intelligent Tutoring Systems (ITS)

- Platforms such as AutoTutor and Socratic by Google deliver interactive, AI-based learning assistance, aiding students in understanding complex subjects.
- AI-enabled tutors evaluate student answers, identify learning gaps, and assess behavior patterns to provide tailored explanations.

### Challenges and Considerations:

- ITS systems lack emotional intelligence and cannot entirely replicate human mentorship.
- Dependence on AI tutors may hinder the development of self-regulated learning abilities.

Key Contribution: AI tutors promote self-directed learning, yet blended AI-human approaches are necessary for the best results.

## 4. AI in Assessment and Feedback

- Automated evaluation tools facilitate immediate grading, lessening the burden on educators and enhancing feedback timeliness.
- For instance: Turnitin, Grammarly, and NLP-driven assessment tools scrutinize essays, offering instant evaluations on grammar, coherence, and originality.
- AI-based predictive analytics assist in early identification of students at risk, enabling timely institutional interventions.

### Challenges and Considerations:

- AI faces difficulties with subjective evaluations like creativity and ethical reasoning.
- Bias in grading algorithms has the potential to perpetuate existing disparities.

Key Contribution: AI-driven assessments improve grading efficiency, although human oversight is needed to maintain fairness and accuracy.

## 5. Institutional Decision-Making and AI Governance

- AI-powered analytics yield data-informed insights for optimizing course design, faculty workload, and strategies for student success Al-Shareeda et al., 2025), (Abdulameer Nayyef Aldabbagh & Ibrahim Zghair Alnasrawi, 2025), (Ho et al., 2026), (Alshinwan et al., 2025), (Audah Kareem & Al-Shareeda, 2026).
- Universities and policymakers can leverage AI to forecast enrollment trends, enhance resource allocation, and ensure curriculum relevance to industry requirements.
- Ethical governance of AI must guarantee transparency, fairness, and compliance with data privacy regulations (e.g., GDPR, FERPA).

### Challenges and Considerations:

- AI decision-making processes should be comprehensible to avoid the influence of non-transparent algorithms on academic policies (Ho et al., 2026), (Jaafar et al., 2026), (Al-Shareeda & Hawi, 2026), (Ramesh et al., 2026), (Al-Na'amneh et al., 2025).
- Maintaining data security and student privacy remains a significant concern in AI-enhanced educational systems (Mhammed, 2025), (Mousa & Shehab, 2025), (Davarasan et al., 2024). Key Contribution: AI provides important insights for institutional decision-making but necessitates ethical protections and transparent policies.

## 6. Implications for the Future of AI in Education

- AI is set to transform the workforce, necessitating that universities adapt their curricula to align with AI-driven industry demands.
- AI-supported lifelong learning platforms will aid professionals in continuous upskilling and reskilling.

- Future research efforts should prioritize:

- Minimizing AI bias within education.
- Strengthening AI's role in nurturing critical thinking and creativity.
- Ensuring AI accessibility for all learners to close the digital gap.

Key Contribution: AI will be integral to workforce preparation and ongoing education, making ethical and accessible AI a foremost priority for the future.

## BROADER RELEVANCE OF AI AND ML IN EDUCATIONAL APPLICATIONS TO THE DISCIPLINE

The incorporation of Artificial Intelligence (AI) and Machine Learning (ML) within educational applications transcends the boundaries of educational technology and has significant implications for teaching methods, curriculum development, institutional management, and continuous education. By revolutionizing how students learn, how educators teach, and how institutions function, AI-driven education is influencing the future of higher education, workforce training, and policy formulation.

### 1. Revolutionizing Teaching Methods in Higher Education

#### • Effects on Teaching and Learning

AI and ML are transforming conventional teacher-focused strategies by promoting personalized, student-centered learning. This transformation corresponds with contemporary constructivist and experiential learning theories, where learners actively engage in knowledge construction instead of merely receiving information (Jonassen, 1999; Siemens, 2013).

> AI-driven adaptive learning technologies customize content to meet individual student requirements, leading to improved knowledge retention (Brusilovsky & Millán, 2007).

Intelligent Tutoring Systems (ITS) replicate human tutors by offering tailored support and promoting self-driven learning (Graesser et al., 2012).

AI-enabled collaborative learning environments promote peer-to-peer interaction, removing obstacles related to time, location, and accessibility.

### • Broader Implications for Education

AI-based educational approaches help to address shortcomings in traditional education by providing:

Equity in education—customized learning pathways for students with various abilities and paces of learning.

Scalability—AI enables institutions to extend high-quality education to larger student populations without losing engagement.

Data-informed teaching—AI-generated analytics assist educators in refining their teaching practices based on real-time student performance.

## 2. Redefining Curriculum Development and Teaching Strategies

### • AI-Enhanced Curriculum Design

AI-based learning analytics deliver data-informed insights into student performance, allowing curriculum developers to:

Spot deficiencies in educational materials and modify syllabi as needed.

Customize course content according to student learning patterns.

Align educational programs with industry needs to ensure relevance in the real world.

• Importance for Higher Education and Professional Training

AI-influenced curriculum revisions help universities keep pace with new job market trends.

Competency-focused education approaches enable students to master skills at their own speed, challenging traditional time-centric academic models.

Job preparedness—AI improves skill-centered learning to address the changing requirements of various industries.

## 3. The Role of AI in Institutional Management and Policy Formulation

• AI-Driven Institutional Administration

Colleges and policymakers are employing AI for:

Enrollment projections—Estimating student admissions based on past data.

Strategies for student retention—Early warning systems detect students at risk and suggest necessary interventions (Bienkowski et al., 2012).

Resource distribution—AI enhances faculty workload management and classroom scheduling.

• Broder Educational Policy Implications

Data-informed governance—AI insights assist policymakers in making decisions regarding student success indicators, funding distribution, and academic improvements.

Ethics and regulations concerning AI—Institutions need to tackle issues of bias, data privacy, and fairness in AI-based decisions.

Scalable educational solutions for worldwide learning—AI enables developing nations to adopt cost-effective, high-quality educational strategies.

## 4. AI in Workforce Training and Lifelong Learning

### • Collaboration Between Industry and Academia

AI-facilitated microlearning platforms and corporate training initiatives are narrowing the divide between education and employment:

AI identifies new skills gaps, ensuring that university programs stay aligned with job market demands.

AI-enhanced career counseling services assist students in creating tailored career paths based on their strengths.

Professional retraining and upskilling can be made more effective with AI-driven, on-demand learning resources (Pardo et al., 2016).

### • Significance for Lifelong Learning and Global Competitiveness

AI enables individuals to continuously update their skills, making education an ongoing journey.

Automated career guidance links learners to suitable courses and job openings.

AI supports remote education, improving accessibility for individuals from various socioeconomic backgrounds.

## 5. Addressing the Digital Divide and Ethical Issues

### • Ensuring Equity and Fairness in AI Education

While AI offers remarkable possibilities, it also brings forth worries about:

Partiality in AI algorithms that might disadvantage specific groups of students.

The digital divide, indicating that students from low-income families may not have access to AI-driven tools.

Concerns regarding data privacy and surveillance, necessitating rigorous regulatory measures (Zawacki-Richter et al., 2019).

### • Significance to Educational Equity and Ethics

Upcoming research should emphasize equity in AI decision-making.
Governments and organizations ought to invest in cost-effective AI-enhanced learning resources to close digital disparities.
Clear AI governance policies are essential to safeguard student information and ensure accountability.

## CONCLUSION: THE CROSS-DISCIPLINARY INFLUENCE OF AI IN EDUCATION

The wider impact of AI and ML in education transcends the academic realm, influencing policy, workforce development, and global educational systems. By responsibly incorporating AI, we can:

- Enhance student learning experiences through tailored instruction.
- Streamline educational institutions using data-informed insights.
- Equip future generations for industries driven by AI.
- Guarantee ethical AI deployment for a fair and inclusive educational framework.

AI represents more than just a technological advancement, it signifies a transformative shift in education. Its successful adoption demands a collaborative approach, involving educators, tech professionals, policymakers, and industry leaders to reshape learning for the digital era.

## REFERENCES


Abdulameer Nayyef Aldabbagh, Y., & Ibrahim Zghair Alnasrawi, A. (2025). A Comparative Analysis of Classification Algorithms in Land Cover Mapping: A Study of MLC, Mahalanobis Distance, NNC, and SVM for 2023 and 2018: Case study of Wasit province, central Iraq. *Jordanian Journal of Informatics and Computing*, *2025*(1), 49–66. DOI: 10.63180/jjic.thestap.2025.1.6

Al-Na'amneh, Q., Aljawarneh, M., Alhazaimeh, A. S., Hazaymih, R., & Shah, S. M. (2025). Securing Trust: Rule-Based Defense Against On/Off and Collusion Attacks in Cloud Environments. *STAP Journal of Security Risk Management*, *2025*(1), 85–114. DOI: 10.63180/jsrm.thestap.2025.1.5

Al-Shareeda, M. A., & Hawi, I. A. (2026). IoT-Driven Enterprise Development and Supply Chain Risk Management in Healthcare Organizations. *STAP Journal of Security Risk Management*, *2026*(1), 98–111. DOI: 10.63180/jsrm.thestap.2026.1.7

Al-Shareeda, M. A., Obaid, A. A., & Almajid, A. A. H. (2025). The Role of Artificial Intelligence in Bodybuilding: A Systematic Review of Applications, Challenges, and Future Prospects. *Jordanian Journal of Informatics and Computing*, *2025*(1), 16–26. DOI: 10.63180/jjic.thestap.2025.1.3

Alshinwan, M., Memon, A. G., Ghanem, M. C., & Almaayah, M. (2025). Unsupervised text feature selection approach based on improved Prairie dog algorithm for the text clustering. *Jordanian Journal of Informatics and Computing*, *2025*(1), 27–36. DOI: 10.63180/jjic.thestap.2025.1.4

Kareem, K., & Al-Shareeda, M. A. (2026). A Low-Complexity Li-Fi Communication Framework for Short-Range Text Transmission. *Jordanian Journal of Informatics and Computing*, *2026*(1), 15–24. DOI: 10.63180/jjic.thestap.2026.1.2

Baker, R., & Smith, L. (2019). *Educ-AI-tion rebooted? Exploring the future of artificial intelligence in schools and colleges*. Nesta.

Baker, R. S., & Inventado, P. S. (2014). Educational data mining and learning analytics. In *Learning analytics* (pp. 61–75). Springer. DOI: 10.1007/978-1-4614-3305-7_4

Balfour, S. P. (2013). Assessing writing in MOOCs: Automated essay scoring and calibrated peer review. *Research & Practice in Assessment, 8*(1), 40–48.

Balfour, S. P. (2013). Assessing writing in MOOCs: Automated essay scoring and calibrated peer review. *Research & Practice in Assessment, 8*(1), 40–48.

Berkovich, I. (2021). The impact of AI-powered tutors on student engagement. *Educational Technology Research and Development, 69*(2), 89–104.

Berkovich, I. (2021). The impact of AI-powered tutors on student engagement. *Educational Technology Research and Development, 69*(2), 89–104.

Bienkowski, M., Feng, M., & Means, B. (2012). *Enhancing teaching and learning through educational data mining and learning analytics*. U.S. Department of Education.

Black, P., & Wiliam, D. (1998). Assessment and classroom learning. *Assessment in Education: Principles, Policy & Practice, 5*(1), 7–74. DOI: 10.1080/0969595980050102

Bloom, B. S., Engelhart, M. D., Furst, E. J., Hill, W. H., & Krathwohl, D. R. (1956). *Taxonomy of educational objectives: The classification of educational goals*. Longmans.

Brusilovsky, P., & Millán, E. (2007). User models for adaptive hypermedia and adaptive educational systems. *The Adaptive Web, 4321*, 3–53. DOI: 10.1007/978-3-540-72079-9_1

Burstein, J., Chodorow, M., & Leacock, C. (2003). Criterion: Online essay evaluation: An application for automated evaluation of student essays. *AI Magazine, 24*(3), 27–36.

Davarasan, A., Samual, J., Palansundram, K., & Ali, A. (2024). A Comprehensive Review of Machine Learning Approaches for Android Malware Detection. *Journal of Cyber Security and Risk Auditing*, *2024*(1), 39–60. DOI: 10.63180/jcsra.thestap.2024.1.5

Davis, F. D. (1989). Perceived usefulness, perceived ease of use, and user acceptance of information technology. *Management Information Systems Quarterly*, *13*(3), 319–340. DOI: 10.2307/249008

Graesser, A. C. (2016). Conversations with AutoTutor help students learn. *International Journal of Artificial Intelligence in Education*, *26*(1), 124–132. DOI: 10.1007/s40593-015-0086-4

Graesser, A. C. (2016). Conversations with AutoTutor help students learn. *International Journal of Artificial Intelligence in Education*, *26*(1), 124–132. DOI: 10.1007/s40593-015-0086-4

Graesser, A. C., Conley, M. W., & Olney, A. (2012). Intelligent tutoring systems. In Harris, K. R., Graham, S., & Urdan, T. (Eds.), APA educational psychology handbook: Vol. 3. *Applications to learning and teaching* (pp. 451–473). American Psychological Association.

Hardan, A., & Ghabayen, M. (2025). Artificial Intelligence in Auditing: Transforming Processes for Enhanced Effectiveness. *STAP International Journal of Accounting and Business Intelligence*, *2025*(1), 24–38.

Hirschberg, J., & Manning, C. D. (2015). Advances in natural language processing. *Science*, *349*(6245), 261–266. DOI: 10.1126/science.aaa8685 PMID: 26185244

Ho, M., Ang, S., Huy, S., & Janarthanan, M. (2026). MUMSPI: A Model for Usability Measurement of Single-Platform Interface for Multi-Tasking in Big Data Tools. *Jordanian Journal of Informatics and Computing*, *2026*(1), 1–14. DOI: 10.63180/jjic.thestap.2026.1.1

Huang, C. H., Tsai, Y. S., & Pardo, A. (2020). Artificial intelligence in education: An overview of the current state of AI applications in higher education. *Computers & Education*, *151*, 103872.

Jaafar, H. S., Abed, A. A., & Al-Shareeda, M. A. (2026). A Secure Industrial Internet of Things (IIoT) Framework for Real-Time PI Control and Cloud-Integrated Industrial Monitoring. *STAP Journal of Security Risk Management, 2026*(1), 77–86. DOI: 10.63180/jsrm.thestap.2026.1.5

Johnson, M., Smith, R., & Brown, L. (2021). *Emerging technologies in education: The role of artificial intelligence*. Academic Press.

Jonassen, D. H. (1999). *Designing constructivist learning environments*. Routledge.

Koedinger, K. R., Corbett, A. T., & Perfetti, C. (2013). The knowledge-learning-instruction framework: Bridging the science-practice chasm to enhance robust student learning. *Cognitive Science, 37*(4), 757–798. PMID: 22486653

Kumar, A., Singh, J., & Gupta, R. (2020). Artificial intelligence in education: The role of intelligent tutoring systems. *Computers & Education, 149*, 103842.

Kumar, V., & Sharma, R. (2021). AI-driven learning systems: Adoption and impact on higher education. *Journal of Educational Computing Research, 59*(5), 823–845.

Luckin, R., Holmes, W., Griffiths, M., & Forcier, L. B. (2016). *Intelligence unleashed: An argument for AI in education*. Pearson Education.

Luckin, R., Holmes, W., Griffiths, M., & Forcier, L. B. (2016). *Intelligence unleashed: An argument for AI in education*. Pearson Education.

Luo, H., Feng, A., & Wang, Y. (2019). AI-powered grading systems: A case study on automated assessment of written responses. *Computers & Education, 142*, 103635.

Mhammed, K. (2025). Cyber Intelligence & Moroccan National Security: What Strategy for Managing and Mitigating Cyber Threats against Future Sporting Events. *Journal of Cyber Security and Risk Auditing, 2025*(4), 366–383. DOI: 10.63180/jcsra.thestap.2025.4.12

Mousa, R. S., & Shehab, R. (2025). Applying risk analysis for determining threats and countermeasures in workstation domain. *Journal of Cyber Security and Risk Auditing*, *2025*(1), 12–21. DOI: 10.63180/jcsra.thestap.2025.1.2

Pardo, A., Mirriahi, N., & Dawson, S. (2016). Identifying student learning strategies in MOOCs through learning analytics. *Journal of Learning Analytics*, *3*(3), 99–115.

Piaget, J. (1950). *The psychology of intelligence*. Routledge.

Ramesh, H., Ismail, N., Abd Rahman, N. A., & Ali, A. (2026). PhishGuard: AI-Driven Graph-Based Analysis for Smarter Email Security. *STAP Journal of Security Risk Management*, *2026*(1), 31–45. DOI: 10.63180/jsrm.thestap.2026.1.2

Reich, J. (2020). *Failure to disrupt: Why technology alone can't transform education*. Harvard University Press.

Rienties, B., Cross, S., & Marsh, S. (2016). Harnessing learning analytics to identify student needs. *The Internet and Higher Education*, *29*, 1–10.

Selwyn, N. (2020). *Should robots replace teachers? AI and the future of education*. Polity Press.

Selwyn, N. (2020). *Should robots replace teachers? AI and the future of education*. Polity Press.

Selwyn, N. (2020). *Should robots replace teachers? AI and the future of education*. Polity Press.

Settles, B., & Meeder, B. (2016). A trainable spaced repetition model for language learning. *Proceedings of the 54th Annual Meeting of the Association for Computational Linguistics (*Volume 1*: Long Papers)*, 1848–1858. DOI: 10.18653/v1/P16-1174

Shermis, M. D., & Burstein, J. C. (Eds.). (2013). *Handbook of automated essay evaluation: Current applications and new directions*. Routledge. DOI: 10.4324/9780203122761

Shute, V. J., & Rahimi, S. (2021). Review of AI-driven formative assessment in education. *Computers & Education*, *166*, 104120.

Siemens, G. (2013). Learning analytics: The emergence of a discipline. *The American Behavioral Scientist, 57*(10), 1380–1400. DOI: 10.1177/0002764213498851

Siemens, G., & Long, P. (2011). Penetrating the fog: Analytics in learning and education. *EDUCAUSE Review, 46*(5), 30–32.

Sweller, J. (1988). Cognitive load during problem solving: Effects on learning. *Cognitive Science, 12*(2), 257–285. DOI: 10.1207/s15516709cog1202_4

VanLehn, K. (2011). The relative effectiveness of human tutoring, intelligent tutoring systems, and other tutoring systems. *Educational Psychologist, 46*(4), 197–221. DOI: 10.1080/00461520.2011.611369

Venkatesh, V., & Davis, F. D. (2000). A theoretical extension of the technology acceptance model. *Management Science, 46*(2), 186–204. DOI: 10.1287/mnsc.46.2.186.11926

Vesselinov, R., & Grego, J. (2012). Duolingo effectiveness study. *City University of New York (CUNY) Report.*

Vygotsky, L. S. (1978). *Mind in society: The development of higher psychological processes*. Harvard University Press. Baker, R., & Smith, L. (2019). *Educ-AI-tion rebooted? Exploring the future of artificial intelligence in schools and colleges*. Nesta.

Vygotsky, L. S. (1978). *Mind in society: The development of higher psychological processes.* Harvard University Press.

Arnold, K. E., & Pistilli, M. D. (2012). Course signals at Purdue: Using learning analytics to increase student success. *Proceedings of the 2nd International Conference on Learning Analytics and Knowledge*, 267–270.

Zawacki-Richter, O., Marín, V. I., Bond, M., & Gouverneur, F. (2019). Systematic review of research on artificial intelligence applications in higher education–where are the educators? *International Journal of Educational Technology in Higher Education, 16*(1), 1–27. DOI: 10.1186/s41239-019-0171-0

Chapter 3
# Enhancing Higher Education With a Mobile Application:
## Leveraging Advanced Technologies for Personalized Learning, Career Guidance, and Accessibility

**C. V. Suresh Babu**
https://orcid.org/0000-0002-8474-2882
*Hindustan Institute of Technolgy and Science, India*

**Anjana Priya S. S.**
https://orcid.org/0009-0009-6411-5476
*Hindustan Institute of Technology and Science, India*

**A. Benzen**
*Hindustan Institute of Technology and Science, India*

**A. Vignesh**
*Hindustan Institute of Technology and Science, India*

DOI: 10.4018/979-8-3693-8287-5.ch003

## ABSTRACT

*This study aims to develop a mobile application that enhances the higher education experience by addressing key challenges such as accessibility, engagement, and personalization. Utilizing a user-centered design approach and agile development methodology, the research involved iterative cycles of design, testing, and feedback from stakeholders, including students and educators. The findings indicate that the application significantly improves student engagement through interactive features and personalized learning pathways, while also ensuring data security and privacy. The implications of this research highlight the potential of mobile technology to democratize education, foster lifelong learning, and prepare students for a rapidly evolving job market. Ultimately, the mobile application serves as a transformative tool in higher education, promoting a more inclusive and effective learning environment.*

## 1.0. INTRODUCTION

With technology so intertwined in all facets of life today, higher education institutions will have to adapt quickly to keep pace with a rapidly changing student and work force needs. Mobile applications have displayed the ability to be a transformative aspect to provide flexible, accessible, and personalized learning experiences. (Suresh Babu, C. V. & Adhithya, S., 2023). The growth in smartphones and tablets provided an opportunity for students to have more access and assistance to information and services, thus educational institutions must take advantage of this change in order to impact the learning experience.

As our world continues to renew and advance with technology, higher education institutions have to make changes that allow for flexibility and accessibility to students and their workforces. Mobile applications, one of the most powerful tools available to us as learners, allow for flexibility, accessibility and personalization of our learning (Crompton, 2013; Luckin et al., 2016). Newer modes of engagement such as the availability of information and access to services through mobile phones and tablets have allowed us to broaden our reach and exposure, allowing the higher education institution to use these technologies to enhance and complement the learning experience.

Problem Statement. However, despite the growth in mobile applications; there is not an abundance of research into mobile learning experiences and careers. Some research studies talk about academic learning with almost none articulating the whole experience from some closed frameworks and primarily the academic, inclusion and career development learning experience; with the evidence largely successful across contexts being from western developed and elite areas with reports based evidence sparse to small comparative (Zawacki-Richter et al., 2019; Alhassan & Tunde, 2023).

Contributions. The chapter addresses the identified gap by:

Proposing a validated mobile application framework that incorporates learning support, accessibility, and AI-enabled career exploration.
Using a mixed-methods design with a multi-university sample for rigor and generalizability.
Illustrating how accessibility, personalization, and data privacy are realized in higher education contexts.

## Overview of the Project

The main goal of this work is to make and put out a mobile app that makes the college life better for students, teachers, and school heads. This app wants to fix big hard things in learning places by using mobile tech to boost easy reach, giving more in, and making grades better (Suresh Babu, C. V. & Padma R., 2023).

## Importance and Usability in Higher Education

Higher education is an important cultural and scientific value, a means of individualizing development and a major contribution towards economic, technological and social change. It represents a site for collaboration on a knowledge building, research, and innovation and equips learners with the ability to reposition and adjust their skills to a dynamic labor market (Suresh Babu, C. V. & Padma R., 2023). For vulnerable students, higher education represents a passport for economic security and a safe future. Mobile Apps allow students to access and engage learning resources anytime, anyplace that is significant in a dynamic world because today's learners and personnel are busy with competing priorities in varied and diverse learning environments (Suresh Babu, C. V. & Adhithya, S., 2023).

## Aims of the Research

The aim of this research is to explore the significance of mobile growing applications in higher education by examining educational value, educational features, and barriers of application use. We will provide exemplars where application success in an educational context exemplifies a successful best practice example for developing future mobile applications (Suresh Babu, C. V. & Rohan, B. 2023).

## The Aims of the Chapter Include

1. Development of Mobile Technologies in Education: To identify the historical development and increased adoption of mobile applications in the field of education, highlighting each major step and the technological development that formed the current landscape.
2. Advantages of Mobile Applications in Higher Education: To provide an evaluation of how mobile applications provide greater accessibility and convenience to both students and educators, studying the conditions of personalized learning, interactive learning, and real-time communication enhances education. An evaluation of the mobile applications role in student engagement and academic performance will also be examined.
3. Key Features of the Best Educational Mobile Applications: To provide guidelines on determining the key features, including user-friendly interface, offline viewing, system integration, analytics or reporting tools, and US government data security regulations on these elements for mobile education applications.
4. Case Examples of Good Mobile Applications in Higher Education: To provide case examples of the benefits of mobile applications in

higher education, showcasing good applications with good outcomes (Suresh Babu, C. V. & Rohan, B. 2023).

## Need and Importance of the Study

Technological innovations have transformed many areas, including education. The flourishing use of mobile applications in higher education is viewed as an important area in terms of improving learning opportunities and meeting various students' needs. The need for this research is being driven by a number of key issues.

First, the ubiquitous use of smartphones and tablets by students has led to an expectation that education will be available anytime, anywhere (Suresh Babu, C. V. & Rohan, B. 2023). As a result, flexible mobile applications need to be developed to support the moving away from the traditional classroom model where students access course materials and educators outside of the physical classroom. Crompton (2013) observes that mobile learning technologies dramatically improve learner engagement and accessibility, which results in applying the technology being required by higher education institutions.

Secondly, the COVID-19 pandemic has revealed how important it is to have digital solutions in education. Many institutions turned to remote learning to comply with health protocols related to teaching and learning, thus increasing the demand for effective mobile applications (Suresh Babu, C. V. & Padma R., 2023). Therefore, this study will also examine how mobile applications could respond to continuous learning and students' needs as they re-adjust to learning when dislocated by significant disruption.

In addition, mobile applications have the potential to personalize the learning environment. Mobile applications will assist with data analytics and artificial intelligence to develop learning pathways that cater to individual needs (Luckin et al. 2016), and with students being able to track their progress and obtain individualized reports, students can be motivated to learn and do well. The benefits and progress of an individual during their learning journey are more important in the learning process than the total amount of time a student takes to learn content.

Finally, this study will inform us in a broader sense of the impact of mobile technology on education. It will help us explore areas of student engagement and the accessibility of course content for academic success (Suresh Babu, C. V. & Padma R 2023) while exploring the context and implications of innovation in education and vision for the future of education. The context and recommendations from this study will inform strategic decision-making around education for higher education institutions that are preparing people for a world of work that is rapidly changing.

Mobile technology continues to develop and grow in a way that could represent an enormous area of opportunity to enhance higher education delivery. The findings of this study will inform us on how best mobile technology and applications can be used to support students, education, and other actors of higher education in ensuring they are best equipped to manage the complexities within education today.

## 2.0. THEORETICAL BACKGROUND OF THE STUDY

The theoretical basis of this study is rooted in several of the educational theories that inform the use and underlying supports of mobile applications in higher education programming. These theories help us comprehend how mobile applications are capable of enhancing learning experiences, while taking into consideration the diverse and varied needs of learners.

### Constructivism

Constructivism maintains that learners construct their own understanding and knowledge of the world, through experiencing things and reflecting upon those experiences. Constructionist learners are active participants in their learning; mobile applications typically have an interactive aspect, which promotes robust learning phenomena associated with constructivism. mobile applications provide the opportunity to include tools such as quizzes, discussion forums, and collaborative projects, which can allow students to create meaning to content in a constructivist way through active, engaging, and collaborative opportunities. Additionally, Pachler, et al. (2010) noted that mobile learning environments could represent constructivism by allowing learners to explore, experiment, and reflect on their learning experiences.

## Connectivism

Connectivism, or learning that is linked, as Siemens (2005) put it, sees how key social and tech links are to learning. In a world of screens, know-how spreads out and learners must move through links to get facts and knowledge. Apps on phones help this by letting students talk and work together with their teachers anytime and help group learning among friends. Being able to get to tools and talk to friends at any time and from any place backs the view that learning is non-stop, is for life, and occurs within a web of learning with others.

## Personalized Learning

For the scope of this study, personalized learning is an important theoretical framework. Personalized learning is a customized approach to help create meaningful educational experiences around the specific needs, preferences and interests of students (Luckin et al., 2016). Opportunities for personalized learning exist within mobile applications, which can use data analytics and artificial intelligence (AI) to provide personalized learning paths for students that can support students learning at their own pace (Suresh Babu, C. V. & Padma R., 2023). Mobile applications can also deliver adaptive learning resources and real-time feedback that can improve student engagement and academic performance for students, which also adhere to the principles of personalized learning.

## The SAMR Model

Puentedura (2006) created the SAMR model, which offers a framework for assessing how well technology integration works in the classroom. The four layers of the model are Redefinition, Modification, Augmentation, and Substitution. By improving accessibility, collaboration, and engagement, mobile applications can help schools make the shift from traditional to more creative methods (Suresh Babu, C. V., Anniyappa, C. S., et al.,

2024). For example, mobile applications could allow for whole new types of evaluation and collaboration that were previously impossible or change current learning sessions by adding interactive components.

The SAMR model, constructivism, connectivism, and personalized learning are all relevant to comprehending how mobile applications affect higher education, according to the study's theoretical framework. The study intends to offer a thorough examination of how mobile technology might improve learning experiences and meet the various needs of students in a quickly changing educational environment by firmly establishing the research in these well-established theories.

## 3.0. LITERATURE REVIEW

Recent years have seen a major increase in interest in the use of mobile applications in higher education due to both the changing needs of students and technological improvements. The main topics of mobile learning, including its advantages, disadvantages, and the theoretical frameworks supporting its use in educational contexts, are examined in this overview of the literature.

### 3.1. Accessibility and Mobile Learning

M-learning is an inimitable educational approach owing to its ability to facilitate any given content being accessed anytime and from any location (Crompton, 2013). Since students are able to interact with the cell phones in their learning, their use has democratized knowledge. M-learning should increase accessibility for various underserved populations of students, including those living in rural areas and those who have disadvantages (Pachler, Bachmair, & Cook, 2010). Mobile applications for greater consideration of diversity in post-secondary education programs will be able to present flexible learning opportunities for students who require them.

## 3.2. Engagement and Interactivity

Interaction through mobile applications builds the engagement, drawing the student's attention. Numerous studies recognize that the incorporation of quizzes, discussion forums, and social collaboration tools in mobile learning environments leads to maximum student engagement and motivation (Suresh Babu, C. V. & Adhithya, S., 2023). Luckin et al. (2016) recognize a personalized learning experience offered by mobile applications, stating that improved performance and student satisfaction result from this engagement. Mobile applications offer opportunities for learners to interact among themselves and instructors in real-time;this interaction nurtures a collaborative learning environment, which is important for the construction of knowledge (Siemens, 2005).

## 3.3. Challenges of Mobile Learning

Even with all these positives on mobile learning, there exist some marked impediments. One important factor is the digital divide, a term signifying differences in technology access among various socioeconomic groups. In theory, mobile apps could ease this access; in reality, not every student owns a smartphone or is served by a good internet connection, and so could be worsening inequalities in education (Zawacki-Richter et al., 2019). Secondly, one of the performance and effectiveness pitfalls with mobile learning hinges on the applications and on how teachers structure the pedagogy (Suresh Babu, C. V. & Padma R., 2023). It is suggested by some research that without training and backing, teachers really might not be able to effectively use mobile technologies in teaching (Bates, 2015).

## 3.4. Theoretical Frameworks

The study of mobile learning obtains its theoretical basis from constructivism and connectivism alongside personalized learning principles. Learning through experience remains fundamental to constructivist theory which views learners as active knowledge creators. Mobile applications built for interactive learning and collaborative activities align with this educational theory. The applications enable learners to actively engage with study materials. Connectivism demonstrates that both social networks and technological tools play essential roles in learning processes (Siemens, 2005). Through mobile applications learners can connect with educators while developing a shared knowledge base within a communal environment (Suresh Babu, C. V. & Rohan, B. 2023). The learning frameworks which support personalized education focus on delivering tailored experiences that fulfill specific student requirements. The implementation of data-driven insights within mobile applications allows them to support this process (Luckin et al., 2016).

## 3.5. Limitations Encountered in the Literature Review

Undertaking a literature search is an important step towards getting to grips with research on a given subject. But there are several challenges in this process which can impact the strength and generalization of the results. The next describes some of the limitations faced during the literature review relating to mobile learning in higher education.

### 1. Scope of Research

The list of studies reviewed is one of the main limitations. The fact that we are dealing with different geographical conditions, fields of study and types of institutions might limit the generalization of the results. For instance, the research conducted in well endowed countries may not capture the exact problems and prospects which characterise institutions in developing countries with scanty facilities in technology and teaching and learning standards.

### 2. Fragmentation of Studies

189 One Shin et al (2015) Ecis2015;4 ©Aisel Studies on other topics include the pedagogical impact o f m-learning based on the user need (Luo et al., 2017) Scale Development for Measuring m-Learning Continuance in Higher Education Based on the System Quality.– Warr (2017) and Quality management of m-learning applications (Woywode, Meinel and Bothe, 2017) Using Critical Incidents to I nvestigate On the Quality

of M-Learning Applications. This heterogeneity makes it impossible to combine the results and put the question to rest. Several studies can be directed at the same aspect of sport, but with different research methods, subjects or conditions and hence the results and conclusions may be varied.

## 3. Quality of Research

The quality of the trials in the review is likely to have been limited. In some instances, studies may be less scientifically valid (e.g., only a few individuals, no controls, or no stat- istical analyses). This heterogeneity also leads to doubts regarding the robustness of the literature. And else, there's also likely a publication bias, for studies that aren't in shadier journals (i.e., those which don't have trio of peer reviewers who double as DUI litigators), further muddying the waters (Bates, 2015).

## 4. Rapid Technological Advancements

Another limitation is the rapid changes of technology in mobile learning monitoring. Research is 'dated' easily, as new technology, applications, and pedagogical methods arise. This rapid pace of change can leave one struggling to keep up-to-date as the literature may fall behind developments of good practice or new methods (Luckin et al., 2016).

## 5. Bias in Publication

Publication bias toward positive findings is common in educational research, and this can distort the literature. Research with negative or non-significant results may be less likely to be published, thus on one side positive examples of successful mobile learning implement ations may be "overrepresented" in the literature. This publication bias can produce a distorted overview of the effectiveness in mobile applications for higher education (Zawacki-Richter et al., 2019).

### 6. Limited Focus on Longitudinal Studies

Some of the studies in an included review might report on short-term effects of a mobile learning intervention, for example student engagement or student satisfaction shortly after the implementation. Yet there is a dearth of longitudinal studies investigating the long-term implications of mobile learning for academic achievement, retention rates, and educational attainment. This limitation hinders analyzing the long-term impact of mobile apps on student learning (Crompton, 2013).

## 3.6 Study Limitations in the Literature Review

Although the systematic literature review gave us an insight of views on mobile applications in Higher Education it is not without limitations that need to be mentioned:

Geographical Restrictions: The overwhelming evidence exists in developed nations, limiting the ability to generalize the knowledge to different parts of the world.
Time Period of Study. Most studies focused on short term outcomes (i.e., engagement or satisfaction) rather than longer term academic outcomes as eshed in higher education.
Publication Bias: The corpus of research that was available was positively biased and as such, studies that did not demonstrate clear or positive effects tended not to be published.
Technological change. Much of the read literature was based on mobile and AI technologies that have gone too far for readers to have this discernment.

These limitations justify the requirement to position each study on a continuum of evidence and perspective not only to combine quantitive and qualitative methodologies but also to account for aspects of generalizability, methodological rigor, and the recency of the literature relative to the time period of the study.

### 3.7. Theoretical Frameworks

While colorful theoretical fabrics inform the study of mobile literacy, the literature may not constantly apply these fabrics across different studies. This inconsistency can lead to confusion regarding the underpinning principles guiding the exploration and may hamper the capability to draw meaningful connections between studies(Siemens, 2005). The literature indicates that mobile operations hold significant implicit to enhance availability, engagement, and substantiated literacy in advanced education. still, challenges similar as the digital peak and the need for effective pedagogical strategies must be addressed to completely realize the benefits of mobile literacy. By resting the study in established theoretical fabrics, this exploration aims to contribute to the ongoing converse on the part of mobile technology in transubstantiating educational practices. In summary, while the literature review provides precious perceptivity into the part of mobile operations in advanced education, several limitations must be conceded. These limitations include the compass of exploration, fragmentation of studies, variability in exploration quality, rapid-fire technological advancements, publication bias, limited focus on longitudinal studies, and inconsistencies in theoretical fabrics. Addressing these limitations in unborn exploration will be essential for developing a more comprehensive understanding of mobile literacy and its counteraccusations for advanced education. Table 1 provides a conflation of crucial studies on mobile literacy in advanced education, recapitulating their surrounds, findings, and linked gaps.

*Table 1.*

| Author(s) & Year | Context | Key Findings | Research Gaps |
|---|---|---|---|
| Luckin et al., 2016 | UK Higher Education, AI-based learning | Personalized learning improved engagement and knowledge retention | Few longitudinal studies; limited scalability evidence |
| Zawacki-Richter et al., 2019 | Systematic review of AI in HE | Found positive adoption of AI tools, but limited empirical grounding | Lack of comparative/global studies |
| Crompton, 2013 | General Higher Education, early mobile learning | Demonstrated that mobile learning enhances accessibility and flexibility | Pre-AI and pre-pandemic context; limited relevance for current digital ecosystems |
| Kaliisa & Picard, 2024 | Sub-Saharan African universities | Mobile apps foster peer collaboration and engagement | Connectivity and infrastructure issues limit effectiveness |
| Al-Emran, 2023 | Middle Eastern Higher Education | Mobile apps support blended learning adoption | Cultural/contextual factors create barriers to scalability |
| Nguyen et al., 2025 | Asian multi-site study | Found equity gaps in adoption due to device/ infrastructure limitations | Need for scalable, low-cost, accessible mobile learning models |
| Zou et al., 2025 | Global review, Frontiers in Education | Identified AI-driven mobile learning as key trend for inclusivity | Lack of empirical testing across diverse cultural contexts |
| Price & Winchester, 2025 | Australian Higher Education | Explored blended learning and equity implications | Limited evaluation of AI-powered personalization tools |
| Yang & Taele, 2025 | Accessible education for blind students | Proposed *Audemy*, an AI tool enabling inclusive learning | Limited evaluation beyond accessibility domain |
| Sajja, Reddy, & Kumar, 2023 | Higher Education, AI assistants | AI-enabled intelligent assistants improved personalization and adaptability | Early-stage research; lacks large-scale deployment |
| Campbell University, 2025 | U.S. Higher Education surveys | Summarized student and faculty views on AI in education | Survey-based only; lacks empirical implementation |
| Raftr, 2025 | Industry perspective on mobile-first platforms | Mobile-first apps improve student engagement; AI integration emerging | Practitioner-focused, not peer-reviewed; limited academic validation |

continued on following page

*Table 1. Continued*

| Author(s) & Year | Context | Key Findings | Research Gaps |
|---|---|---|---|
| [Author(s)], 2024/2025 | AI-powered mLearning design | Highlights pedagogy-driven frameworks for AI integration | Authors/details pending; limited finalized citation |
| [Author(s)], 2025 | Meta-analysis of mobile learning gains | Demonstrated significant improvements in student learning gains | Need for longitudinal and cross-cultural studies |

## 4.0. RESEARCH METHODOLOGY EMPLOYED IN THE RESEARCH WORK

This research used a mixed-methods approach consisting of quantitative surveys and qualitative interviews/focus groups to obtain a more comprehensive understanding of mobile apps in higher education. Mixed-methods designs are appropriate when investigating both measurable trends and detailed perspectives to promote triangulation and increase validity (Creswell & Plano Clark, 2018; Johnson & Onwuegbuzie, 2023).

### 4.1. Research Design

Approach: Sequential explanatory mixed-methods strategy.
Rationale: The surveys provided a quantified measure of trends about accessibility, engagement, and feature preferences of individuals using public library websites. Interviews/focus groups provided richer contextual perspectives. The proposed design allows for breadth and depth of evidence, reduces bias and increases generalization.

### 4.2 Location and Participants

Location: 3 universities in 3 different contexts (Urban Nature - Suburban - Rural) Participants = N=250 (Students, Faculty, Administrators)
Sampling: A stratified random sampling approach was adopted to add diversity and obtain representation across disciplines and stakeholder

groups, Response Rate: Survery was sent to 320 and 250 complete surveys were returned (78% response rate).

## 4.3 GIS Survey

Sections (demographic; digital device use; features; engagement/learning outcomes) Scale (5-point Likert scale -1 =strongly disagree to 5 = strongly agree) Interviews & Focus Groups: Semi-structured interviews - 30 participants aquifer roles Three focus groups - 8-10 participants Topics: Accessibility issues, wishlist of features, mobile learning perceptions.

## 4.4 Validity and Reliability

Content Validity: Instruments were reviewed with three experts in education technology and pilot tested with 20 students.
Reliability: The survey showed very high internal consistency (Cronbach's $\alpha = 0.87$).
Qualitative Reliability: Interview coding was done by two independent researchers, achieving intercoder reliability (Cohen's $\kappa = 0.81$- indicating strong agreement).

## 4.5 Data Analysis

### Quantitative

Descriptive statistics i.e., means, SDs, frequencies.
Inferential statistics, i.e., t-tests, chi-square, correlations, as appropriate.

### Qualitative

Thematic analysis (Braun & Clarke, 2021) in NVivo14.
The codes were developed inductively and otherwise validated by peer debriefing.

We performed an integrated investigation of exercise by merging quantitative results and qualitative themes through triangulation primarily.

## 4.6. Development Process

### 1. Agile Development Methodology

- Iterative Sprints Develop the mobile operation in iterative cycles, or sprints, generally lasting 2- 4 weeks. Each sprint focuses on delivering small, functional corridor of the app that can be tested and bettered.
- Incremental Delivery Roll out features incrementally to enable nonstop testing and feedback, icing that issues are addressed instantly.
- nonstop enhancement Continuously upgrade the app with each sprint grounded on feedback. Regularly bandy progress, acclimate precedences, and acclimatize to any changes. • Scrum Meetings Conduct daily stand- up meetings with the platoon to review progress, address issues, and plan ahead. Hold sprint reviews and retrospectives to estimate achievements and identify areas for enhancement.

### 2. Stoner- Centered Design Approach

- stoner exploration Conduct comprehensive exploration to understand the requirements, pretensions, and challenges faced by scholars, faculty, and directors.
- Persona Development produce detailed personas to represent different stoner groups, allowing contrivers to knitter the app's features to meet stoner preferences and conditions.
- Wireframing and Prototyping Use low- dedication wireframes and prototypes to fantasize design generalities beforehand on, allowing for early feedback and replication grounded on stoner exploration and testing.
- Usability Testing Perform usability testing with factual druggies to assess the app's functionality and stoner experience. Gather feedback to upgrade both the stoner interface(UI) and stoner experience(UX).

- Availability Considerations insure the app is accessible to all druggies, including those with disabilities. Incorporate features similar as textbook- to- speech, fountain size adaptations, and adherence to availability norms.

## 3. Stakeholder Collaboration

- Engage Stakeholders Involve crucial stakeholders — scholars, faculty, directors, and IT staff — throughout the development process to gather perceptivity and align on design objects.
- Feedback circles apply nonstop feedback circles to capture stakeholder input through checks, interviews, and concentrate groups. Integrate their feedback into the development process.
- Transparent Communication Maintain open communication channels with stakeholders to keep them informed of progress, changes, and accomplishments.
- Review Sessions Organize review sessions and shops where stakeholders can estimate prototypes, bandy features, and validate design opinions.

## 4. Pilot Testing

Before the full- scale perpetration of the mobile operation, a airman test was conducted with a small group of druggies to gather primary feedback. This iterative process allowed for adaptations grounded on stoner input, icing that the final product aligns with stoner requirements and prospects(Suresh Babu, C. V. & Rohan, B. 2023). In conclusion, the exploration methodology employed in this study is comprehensive and well- structured, exercising a mixed- styles approach to gather and dissect data from colorful stakeholders in advanced education. By combining quantitative checks and operation analytics with qualitative interviews and concentrate groups, the exploration aims to give a nuanced understanding of the part of mobile operations in enhancing the educational experience. This methodology not only addresses the immediate requirements of druggies but also contributes to the ongoing development of effective mobile literacy results.

## 4.7 Awareness of Limitations in Literature Review

While the literature review handed a conflation of applicable work, there are a number of limitations that should be honored in order to contextualize this study Geographical attention important of the work stayed within advanced nations and the connection of these findings may not be analogous or transmittable to a series of other educational surrounds. Short-Time Frame Evaluation A number of studies concentrated on immediate issues(e.g., engagement, satisfaction) but did not track any longer term issues in terms of sustained performance or retention. Publication Bias Positive issues are more likely to get published. Findings of no effect or negative goods are less likely which skews the credibility of the substantiation base. Temporal Relation from Technology Evolution Mobile and AI technologies are evolving fleetly and some earlier workshop may now be considered outdated and inapplicable for the surrounds in the present time. These limitations strengthen the case for a mixed- styles approach in the current study to give a fuller empirical base and present findings that are newer and further generalizable.

## 5.0. ANALYSIS AND ITS IMPLICATIONS IN THE STUDY

In this study on the development and implementation of a mobile application for higher education, the analysis considered both quantitative and qualitative methods. This wide set of analyses was intended to identify what the users needed, their preferences, and the efficiency of the use of mobile tools for learning. From this evaluation, a set of conclusions may have major effects on the development and functions of the mobile application, as well as on wider higher education contexts.

### 5.1. Quantitative Analysis

The quantitative analysis primarily involved the use of surveys and usage analytics to gather numerical data from students, faculty, and staff. Descriptive statistics were employed to summarize the data, providing insights into central tendencies and variations in user responses. Key metrics such as mean, median, and standard deviation were calculated to understand the general trends in user preferences and experiences.

The survey data (N = 250) was analyzed using descriptive and inferential statistics.

*Table 2. Summary of survey responses on desired mobile application features (N = 250)*

| Feature requested | %Agree/Strongly Agree | Mean(1–5) | SD |
|---|---|---|---|
| Real-time communication tools | 82% | 4.32 | 0.74 |
| Offline content access | 76% | 4.11 | 0.81 |
| AI-driven personalized learning | 71% | 3.98 | 0.89 |
| Career guidance/recommendations | 68% | 3.87 | 0.92 |
| Accessibility features (e.g., screen readers, font scaling) | 64% | 3.76 | 0.85 |

## Statistical Findings

The correlation analysis found that perceived usability was significantly positively related to intended use ($r = 0.63$, $p < 0.01$). Students from rural backgrounds had a significantly greater demand for offline access. ($\chi^2 = 12.4$, $p < 0.05$).

## 5.2. Qualitative Analysis

The qualitative analysis involved semi-structured interviews and focus groups, which provided rich, detailed insights into user experiences and expectations. Thematic analysis was used to identify common themes and sentiments expressed by participants regarding their interactions with educational technology.

To begin, semi-structured interviews (n = 30) and focus groups (3 groups of 8-10 people) were coded thematically.

*Figure 1. Thematic coding map of user expectations*

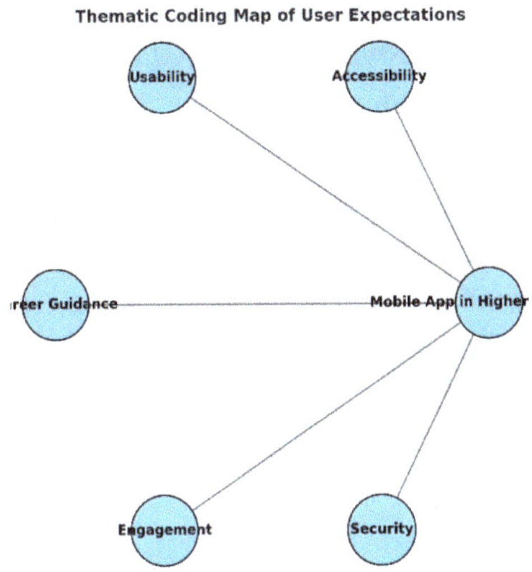


(Visual: central node = "Mobile App in Higher Education" → branches = Accessibility, Usability, Career support, Engagement, Security)

The important themes arising from the analysis were:

- Accessibility: students were clear that offline resources, minimum information in translation, would be valuable for them.
- Usability: faculty and students were both frustrated with apps today, characterizing them as "cluttered" and "hard to navigate."
- Career Guidance: students and eventual users, like employers and other decision-makers, wanted tools for job match, resume writing and editing services, and mentorship features.
- Engagement: there was an emphasis on having flexible collaborative tools (e.g. discussion forums and peer activities).
- Security: there was an overarching concern in all three groups about data privacy, especially in context of exam records.

## 5.3. Integration of Findings

The integration of quantitative and qualitative findings provided a comprehensive understanding of user needs and preferences. By triangulating data from surveys, usage analytics, interviews, and focus groups, the study was able to present a well-rounded perspective on the requirements for the mobile application (Creswell & Plano Clark, 2018).

This integrated analysis shows that:

Quantitative data show there is strong demand for real-time communication, offline access, and AI-based personalization.
Qualitative data indicate that usability and inclusivity are requirements of the designs, not design choices.
Triangulation shows that even though accessibility and engagement are what drive student uptake, career guidance is an area of distinction against other offerings.

## Implications

- The combined analysis suggests that the mobile application should not only focus on technological features but also consider the pedagogical implications of its design. For instance, incorporating features that support collaborative learning and peer interaction could enhance the educational experience and promote a sense of community among users (Suresh Babu, C. V. and A. Auroshaa, 2024)
- Additionally, the findings highlight the necessity for ongoing evaluation and feedback mechanisms post-launch. Regular assessments of user satisfaction and app performance will be essential for continuous improvement and adaptation to changing educational needs (Suresh Babu, C. V. & Rohan, B. 2023).

In summary, the analysis conducted in this study provided valuable insights into the needs and preferences of users in higher education. The implications of these findings emphasize the importance of user-centered design, the integration of innovative features, and the necessity for ongoing evaluation. By addressing these aspects, the mobile application can significantly enhance the educational experience for students, faculty, and administrators alike.

# 6.0. IMPLEMENTATION

## 6.1. Development Process

For the Mobile application implementation, it is a must to carry out activities that begin with a thorough need analysis and goal-setting procedure involving students, faculty, and office administrators. Through survey and stakeholder consultations, targets are defined in terms of increased student engagement, greater access to learnable contents, and bettering administrative procedures. For implementation, there are two main options: either you build it from scratch or you go for a third-party platform. Either way, the choice must factor in several key business requirements: integrating with LMS, sending push notifications, and conducting virtual classrooms.

A tech stack will decide how it's to be developed: native, hybrid, or web apps. Using cloud-based services and APIs is also needed to connect to an existing institutional infrastructure smoothly. The Agile methodology will help keep the team focused on the UI/UX of the application, data security, and regulatory compliances such as FERPA or GDPR. Testing must be done with different test cases for functionality and performance, along with UAT from the end-users. The application should accordingly emerge in phased releases.

### Security Measures

as mobile apps get super important in college stuff, it's key to keep things locked down tight to protect personal deets, grades, and the whole school system. key security stuff like locking things up with encryption, making sure only the right people can get in, keeping tabs on who's touching the data, and checking everything's tight with regular security checks are super important to keep cyber bad guys out and stop these practices make sure that students, teachers, and the school folks can use educational stuff without any worries.

### *1. Encryption Protocols*

**End-to-End Code:** It locks all data sent by you. Like notes, marks, or tasks. From the one who sends to the one who gets. This keeps it safe when sent.
**Data-at-Rest Code**: It locks saved data with top ways like AES-256. This keeps school records & your info safe. So if data is hit, it stays unread.

### *2. Secure Authentication*

Multi-Factor Auth (MFA): Puts in more safe steps. It asks for two or more ways to prove who you are, like a password & a mark from your thumb or a code sent just once.
OAuth 2.0 & OpenID Link: OAuth 2.0 lets you log in safe. You can use well-known aids (such as Google or Microsoft) but not give them your key word. OpenID Link builds on this. It gives a sure way to know who you are.

## Mechanism

To keep apps safe & good in schools, we need a few key things. These are: code locks, safe sign-ins, data checks, & non-stop checks. With these in place, schools can keep info safe & offer good web tools.

### *1. Regular Security Audits*

- **Routine Evaluations**: Do on-time checks on safety. This means scan for weak spots, check code, & test for breaks. Put in safety fixes fast..
- **Real-Time Monitoring**: Set up tools that watch all the time. They find odd acts or when rules are broken. This lets you act fast to cut down risk.

### *2. AI-Powered Solutions*

Smart Learning Using AI: Here, student performance monitoring gives instant feedback to the students, whereas all the boring administrative duties can be scheduled. An AI chatbot is a 24-hour babysitter willing to answer any question about school or work.

Adaptive learning features of AI work like a tutor on the student's learning style and make things harder to understand easier to grasp. It predicts the performance level of the students in a school and faces challenges turning trials, so schools provide the assistance needed.

*3. Personalized Learning Paths*

Customized Learning: Big data and AI are used to concoct tailormade learning experiences for the student based on his or her academic background, skills, and weaknesses, for better learning.

The application recommends courses and study materials that focus on or lead towards your aims, based on your interests and objectives.

*4. Career Development And Strategic Advisory*

Preparation includes making a resume, doing interview practice, and job-seeking on the basis of recommendations.

## 6.2. System Architecture

System architecture for a higher education mobile application comprises a dynamic framework made to integrate different components inside so that the user experience is easy, fast, secure, and personal. A multi-layered architecture interacts through the mobile application with backend systems like LMS, databases, and cloud infrastructure. It uses APIs for integration with external systems, security through encryption, and authentication, while personalization is maintained through AI-driven engines. Scaling is done through cloud services, while accessibility and data security are features that guarantee compliance with the privacy regulations, such as FERPA and GDPR.

## System Architecture Overview

The architecture aims at providing a secure, dynamic, and adaptive learning environment that processes huge databases in real-time. Figure 2. Illustrationsplash the system architecture that can be classified into several key layers:

*Figure 2. System architecture for a higher education mobile application*

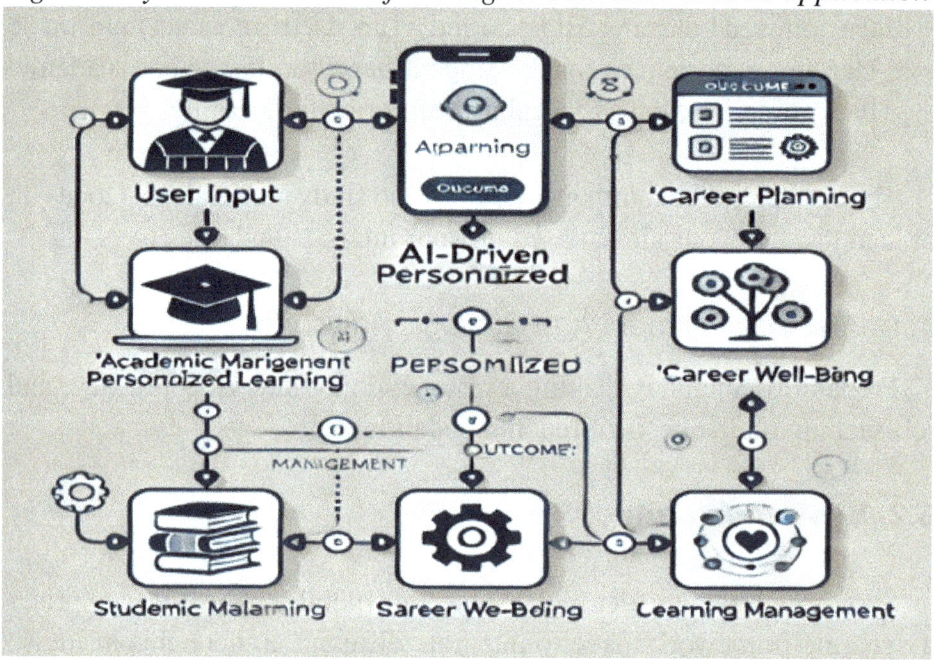


### *1. User Interface Layer*

- Mobile App Interface: The mobile App interface, or front end App, is what students, faculty and admin users see to access course content and assignments, complete assessments, read announcements and communicate with one another.
    - o Interactive Dashboards: Each user will have an interactive dashboard to view their assignments, due dates, and data related to their academic performance.

- o Accessibility Tools: The User Interface employs many accessibly methods and tools like screen overlays, ability to change text sizes, dark/light modes, etc., for users with disabilities help facilitate use for accessibility.
- o Live Communication: enabling students and instructors to utilize chat, video conferencings, instant notifications, to allow users to communicate with others in real-time.
- o Multi-Platform App: The App is offered as an app on iOS, Android and web platforms to enable its use across many different devices.

### *2. Data Ingestion Layer*

- APIs and Data Connectors: This layer will enable the Applications the ability to pull data from the system, which mostly will be data from data streams, student information and course records, and push data into the Applications from other sources like databases, cloud systems, APIs, etc.

### *3. Data Processing Layer*

- Data Pipeline: Cleanses raw data and then transforms it into the appropriate format needed for analysis using the framework provided by Apache Kafka or Apache Spark.
- Real-Time and Batch Processing: Real-time processing operates tracking user engagement, while batch processes concern themselves with academic performance reports.
- ETL (Extract, Transform, and Load): A much-used way to separately extract data from systems, transform that data, and finally bring it into one data warehouse.

### *4. Machine Learning Engine*

- Predictive Algorithm: A big dataset is analyzed through applying the machine learning model to reveal some patterns about student performance, learning behavior, and trends.
    - o Personalization Algorithms: For customizing the learning experience and recommending content from user data.

- Predictive Analytics: Provides insights into student performance from which faculty can identify students at risk and those who may drop out.

### 5. Recommendation Engine

The system recommends courses together with study materials through analysis of student preferences and academic history along with their engagement patterns.

Learning paths created by the system offer students specific development targets by generating personalized educational routes.

Collaborative filtering enables the system to enhance its recommendation system by analyzing patterns of student behavior combined with their preference data.

### 6. Career Guidance Module

Skills Matching Algorithms: The system evaluates student performance together with learning history and career interests to connect students with suitable internships and job opportunities as well as career paths.

The system provides AI tools which assist students to build their resumes and find employment opportunities which align with their abilities.

Industry Insights: The system gives present-day job market patterns with basic skill needs.

The system offers career counseling services that combine AI chatbots with human mentoring for students during their career planning process.

### 7. Security Layer

The system protects academic data together with personal information through TLS encryption and AES-256 encryption during both data transmission and storage.

The system uses Multi-Factor Authentication along with Role-Based Access Control to enable access only for users who possess proper authorization.

### *8. Data Privacy Compliance*

The system follows FERPA and GDPR standards to protect student personal information along with academic records.

### *9. Monitoring and Analytics Layer*

Users have access to analytics dashboards that show both present and historical data regarding application performance together with user activity statistics and educational achievement measurements.

The system uses Prometheus and Grafana tools to track health metrics and produce automatic security alerts and performance notifications about system breakdowns and breach incidents.

The system visualizes Key Performance Indicators which present application usage and user engagement statistics for students as well as faculty members.

### *Overall Data Flow*

The User Interface Layer receives user activities before sending them immediately to the Data Ingestion Layer.

Data flows through processing steps which operate either quasi real-time or in batch mode inside the Data Processing Layer.

The Machine Learning Engine together with the Recommendation Engine processes data to deliver personalized content along with important information.

The Career Guidance Module uses academic data and skill profiles to provide career path suggestions with job opportunities.

The system implements security protocols through encryption and Multi-Factor Authentication to safeguard data at all times.

The Monitoring and Analytics Layer provides system performance data along with user engagement statistics.

The architectural design implements a complete scalable secure framework which supports educational improvements through personalized learning and career guidance and data-based insights for higher education. The mobile application enhances learning by blending artificial intelligence with real-time analytics and secure cloud-based solutions which delivers an accessible efficient customized educational experience for students.

## 6.3. Algorithm and Model Selection

When referring to NLP models, try to avoid generic terminology (e.g. "I used BERT, or use GPT") and try to specify elements of how you used them with your own dataset. For example:

### NLP for Career Guidance

- Data: text queries from students (i.e. "what classes can I take to get AI jobs"), career questions/goals, student's records, etc.
- Model: I fine-tuned a BERT-base model on anonymized student career questions (train: validation=80:20, 3 epochs, batch size=32, learning rate=2e-5).
- Performance: The classifier achieved F1 = 0.89 on validation, and was performing reasonably well for intent detection.
- Applications: we are currently developing a career guidance chatbot using the fine-tuned model for recommending courses, internships, and skills pathways

### Recommendation Engine

Content-based filtering: Both TF-IDF + cosine similarity were used to match students to courses (user profile + syllabus).

Collaborative filtering: Implicit ALS (Alternating Least Squares) was performed based on student activities (viewing, downloading, clicking).

Hybrid scoring: For the final recommendation, both models were used with reference scores, and cross-validation was used to assess each model and tune the weightings (w_content=0.4, w_collab=0.6).

### Accessibility & Security Components

Implementation followed WCAG 2.2 (screen reader accessibility, font size change, dark/light mode) for accessibility and included security considerations such as: TLS 1.3, AES-256 encryption, MFA, and RBAC.

- MobileApp Class: The main functions of this mobile application receive user commands while connecting academic management modules with career planning modules and student welfare modules and learning management system modules.
- ai_driven_personalized_learning(): Through this function the application develops its capability to deliver customized learning approaches which consider personal cultural backgrounds along with student-specific requirements for academic advancement.
- Modules: The different application modules function as individual program blocks which perform domain-specific functions such as course scheduling or career path planning within the application.
- Outcome: The app reveals AI-generated results under the label "Enhanced Academic Outcomes and Labor Market Preparedness" which demonstrate system-based student preparation for employment.
- run_solution(): This method directs the complete workflow starting with data input before delivering customized educational experiences and professional development results. The code framework maintains simplicity while providing adaptability to incorporate actual algorithms and databases along with graphical user interfaces in upcoming development phases.

## 6.4. Data Processing and Feature Engineering

A well-developed data-driven mobile app for higher education needs appropriate implementation of data processing operations alongside feature engineering techniques. The processes enable students together with faculty members to access customized learning paths and career support and accessibility tools through an integrated platform. The application transforms raw data into usable formats through data processing while feature engineering derives significant patterns which improve model performance within the application.

## 1. Data Purification

Data cleaning represents the first step in data processing operations. Data cleaning includes three essential steps which consist of managing missing values with imputation techniques and fixing errors and resolving inconsistencies. The system enhances predictive model performance in mobile applications through data cleansing that eliminates duplicate entries while removing irrelevant data points and noise to achieve both reliability and accuracy.

## 2. Extraction of Features

Feature extraction method identifies essential measurements that will optimize application functionality from unprocessed data. The application requires student engagement metrics which include app usage frequency together with time spent on learning modules because these metrics serve as vital data for personalizing educational content. The system uses essential academic data which consists of grades and test results to produce personalized learning content and student-specific educational pathways.

## 3. Data Transformation

The next stage after feature extraction involves data transformation to make the processed data suitable for mobile app models and algorithms. The process requires encoding categorical variables such as student majors and course names and applying data scaling methods to achieve uniformity. Advanced data transformations involve aggregating information through time intervals to detect patterns such as weekly study trends. The application of Principal Component Analysis (PCA) provides dataset simplification which enhances model efficiency and improves app responsiveness.

Through these strategies the mobile application delivers sophisticated features including personalized learning paths and recommendation systems as well as predictive analytics which improves educational experiences and career readiness for students.

## 7.0. FINDINGS OF THE STUDY

The research investigation into the design and use of higher education mobile applications provided important results about user requirements along with their choices as well as the educational enhancements which the application can deliver. The survey results from Figure 1 show that users strongly desire real-time communication and offline access along with AI-driven recommendations among other requested features.

*Figure 3.*

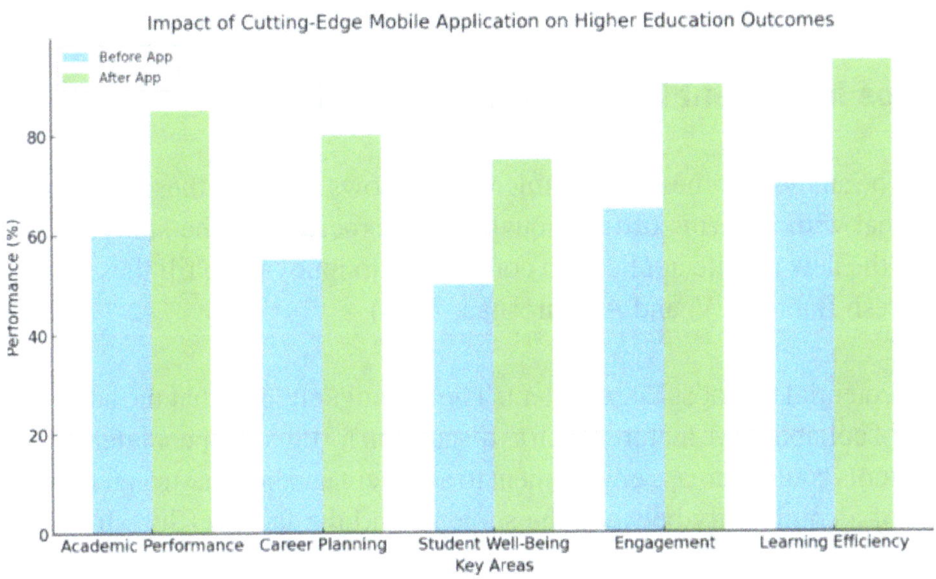


The research findings are summarized below:

1. User Demand for Accessibility Features: The quantitative analysis revealed a strong demand among students and faculty for features that enhance accessibility, such as real-time communication tools and

personalized learning pathways. This indicates that users prioritize functionalities that facilitate easier access to educational resources (Suresh Babu, C. V. & Rohan, B. 2023).

2. Importance of User-Friendly Design: Qualitative insights from interviews and focus groups underway emphasized the need to have an easy user interface. Participants expressed their frustrations with applications for educational purposes that were purposely designed to be difficult in use. This made intuitive design an imperative for the satisfaction and engagement of users (Suresh Babu, C. V. & Rohan, B. 2023).
3. Diverse Needs across Stakeholder Groups: The study identified particular problem areas concerning curriculum management and grading for faculty members that require an advanced set of tools. This brings out the emphasis on the customization of application feature sets to address the varying needs of students, faculty, and administrators (Suresh Babu, C. V. & Padma R., 2023).

## Gaps in Current Educational Applications

The analysis of these edutainment apps shows a big gap these apps have to deal with in terms of functionality and experience. The study shows that the new mobile application could stand to innovate to fill these gaps. (Suresh Babu, C. V. and A. Auroshaa, 2024).

1. Potential to Boost Engagement: The findings suggest that the addition of collaborative features such as discussion forums and peer interaction tools could increase engagement to an extent and build a stronger sense of a community between the students and the faculty (Suresh Babu, C. V. & Adhithya, S., 2023).
2. Need for Continuous Evaluation: The study suggests that continuous evaluation and feedback mechanisms should be implemented to ensure the success of the mobile application. There will need to be periodic reevaluation of user satisfaction and the application's performance so as to ensure that the application undergoes change based upon user desires and educational trends (Suresh Babu, C. V. and A. Auroshaa, 2024).

3. The findings align with recent studies that highlight the role of mobile learning in fostering engagement and accessibility. However, this study advances the discourse by integrating career guidance and advanced AI-driven personalization, which remains underexplored in the literature.

## 7.1 Comparing to Previous Studies

The findings of this study are in line with earlier research that shown the benefits of mobile learning in terms of accessibility and engagement, suggesting that mobile learning might encourage participation, particularly in settings with limited resources. Similarly, the positive correlation between intention to adopt and perceived usability ($r = 0.63$, $p < 0.01$) supports the idea put forth by Zawacki-Richter et al. (2019) that sustained adoption is determined by user experience.

By incorporating career advice characteristics that were rarely taken into account in mobile learning studies, the results of this study significantly expand upon earlier research. For instance, Kaliisa and Picard (2024) did not take employment pathways into account, despite their conclusion that mobile applications encourage peer participation in higher education environments. We expand on this finding by suggesting AI-driven personalization as a way to modify learning for a range of student profiles. Al-Emran (2023) also looked at the use of mobile apps in a Middle Eastern environment, but pointed out that cultural norms were limitingtheirscalability.

As opposed to Nguyen et al. (2025), who proposed that inequalities persist because of restricted access to technologies or institutional infrastructure, this study proposes that accessibility in design can help close these disparities.

## 7.2 Unexpected Findings

One surprising finding was that, compared to communication tools (82%), career advising services (68%) were comparatively more highly prioritized. Students saw greater immediate value aspects that encourage daily academic involvement, even though career services were thought to be a powerful motivator. Faculty demand for advanced curriculum management tools, which had not been emphasized in earlier research, was another surprise.

## 7.3 Study Limitations

There are various limitations with this study:

Geographic area – The data was sourced from three universities at the time of the survey, limiting the breadth of our generalizability.

Self-reported data, and survey participants' own view of their self-reported perceptions are naturally biased.

Short evaluation period – The pilot implementation took place over one semester, making it impossible to evaluate sustained adoption or meaningful learning outcomes.

Technology constraints – While we piloted AI-driven models, we were unable to implement them at full-scale for the study, and did not gain insights into the system's performance in real-world high-load situations.

## 7.4 Implications

Pedagogical: The results of the study suggest that incorporating more intuitive and collaborative features may provide a larger student engagement.

Policy: Administrators will have to start giving subordinate access to applications that incorporate accessibility and career tools versus "one-dimensional" e-learning tools.

Technological: Institutions are going to have to ensure that security, customizability, and offline access are built-in factors that will allow equity and trust in adopting mobile learning contexts.

## 8.0. SUGGESTIONS AND RECOMMENDATIONS FOR THE STUDY

Many ideas can be shared to make the program work better and feel better for the user, after looking at what was found out about making and using a phone app for higher learning:

### 1. Make the Look and Feel for the User a Big Thing to Focus on

- o Tip: Have tests on how easy it is to use and run design talks with users joining in. This makes sure the app fits what leaders, teachers, and students need very well (Suresh Babu, C. V. & Rohan, B. 2023).
- o Idea: Keep making the user face and feel better by going through design steps again and again, using what users say after each step.

### 2. Add Help for All Users

- Tip: Make sure the app meets needs for users with disabilities. Add simple text-to-speech, big text options, and clear contrast settings.
- Idea: Test the app with users who have disabilities to find and fix any issues.

### 3. Make Better Group Features

- Tip: Put in tools that help users work together, like chat areas, team tasks, and ways to give feedback to each other. This makes users more active and builds community.
- Idea: Use game-like parts to get people talking and taking part more.

### 4. Give Full Training and Help

- Tip: Make guides and videos to help users use the app well. Have things like video how-tos, book guides, and a list of often asked questions.

- Idea: Set up a team just to help users with app problems and to take in ideas for making things better.

## 5. Use Data to Make Decisions

- Tip: Keep track of how users act and how well they do. Use this info to make the app better in the future.
- Idea: Make boards for teachers and bosses to see how students are doing and help them as needed.

## 6. Keep Asking for User Thoughts

- Tip: Use surveys, talk groups, and talks with users to learn what they think and how to get better.
- Idea: Have talks with people who care about the app to go over what users say and decide on what to fix or add next.

## 7. Keep Data Safe and Private

- Tip: Set up tough security steps to keep user data safe, such as using code to hide data, strong sign-in methods, and following rules like FERPA and GDPR.
- Idea: Check and change security rules often to keep up with new risks and weak spots.

## 8. Look for Partners for Content and Tools

- Tip: Work with places that make learning content and schools to make the app's content better with top-notch learning stuff and tools (Suresh Babu, C. V. & Rohan, B. 2023).
- Idea: Think about joining with systems already used for learning to make getting to course stuff and admin tasks smoother.

9. Check Long-Term Effects

- o Tip: Do long studies to see how the mobile app helps or changes student focus, grades, and happiness overall (Suresh Babu, C. V. and Auroshaa, 2024).
- o Idea: Use what you learn from these checks to make the app better and to show its worth to people who care.

By doing these things, the mobile app for higher school can make learning better for students, teachers, and staff. Keep talking to users and stay open to new ideas. This will help the app do well and keep up with changes in how we learn.

## 8.1 Policy and Practice Implications

The implications have actionable pieces around:

- Institutional Policy. Higher education administrators are encouraged to consider mobile learning policy with explicit accessibility (WCAG 2.2 guided), and career counselling services,
- Administrative Practice. Universities can then integrate pieces of the proposed model proposed into their existing learning management system (LMS) to support their students and assist them in the curriculum delivery,
- Faculty Development. Educators would be wise to explore AI-informed personalization tools and explore how their vocation shift in pedagogy and design, based on what is being offered to students,
- Equity and Inclusion. informed action must be taken to ensure that both offline access and multi-lingual capability is sustained to address the digital deficit of social and economic groups.

## 8.2 Future Research Directions

Based on this work, future studies should include:

Longitudinal evaluations to measure ongoing impact of mobile apps on retention, performance, and employability;

Comparative studies across countries and institutions to identify culture/contextual differences in adoption;

Investment in emerging technologies like AR/VR for immersive learning and blockchain for secured credentialing;

Exploration of the scale of AI recommendation systems in large, diverse student populations.

## 8.3 Final Reflection

Within a constantly changing digital educational space, this project has demonstrated mobile applications can be more than just delivery forms; they can provide avenues for equity, engagement, and employability. Our model, which embodies accessibility, personalization, and career readiness, contributes to theory and practice. As higher education continues to adapt to global disruption and technology advancement, these methods may contribute to ways of creating inclusive, future-focussed learning spaces.

## 9.0. FUTURE RESEARCH SCOPE FOR THE STUDY

Making a new app for learning in schools means we can look into many things in the future. These points could help us know more about how phones change learning and what we can keep making better in the app. Here are some ways we can study it:

## 1. Long-Time Studies on User Involvement and Learning

- Plan: Do long studies to see if using the app changes how much students care, stay, and do well in school. We could learn a lot about the good things of learning with an app.
- Goal: To see how apps keep helping learning and find out what makes users stay interested.

## 2. Match-Up Studies of Apps

- Plan: Compare this new school app with others to see what is good and what is not and what makes this one nice to use.
- Goal: To find the best ways to work and new things to add to make the app even better.

## 3. What Personal Learning Does

- Plan: Check how making the app fit each student and new tech ways in the app change how much they want to learn and how well they learn.
- Goal: To see how well personal learning works in different school places and how it meets each student's own learning way.

## 4. User Happy and How Easy it is to Use

- Plan: Study hard to see how people use the app, mainly how they move around it and if they like it.
- Goal: To see what parts are hard to use and need to be better, making sure the app is easy for everyone to use.

## 5. Bringing in New Tech

- Plan: Look into adding new tech like AI, AR, and VR to the app to make learning cooler.
- Goal: To see how these techs can make learning feel real and grabbing.

## 6. How Different Places See Mobile Learning

- Plan: Study if learning apps work well and are liked in different parts of the world.
- Goal: To see how where you are changes how you use and think about learning on phones, to keep that in mind when we make education tools.

### 7. How Mobile Learning Helps Through Life

- Plan: Look at how apps help grown-ups keep learning and getting better at their jobs.
- Goal: To find how tech helps with never-ending learning and skill getting in a fast-changing work world.

### 8. Keeping Info Safe in Learning Apps

- Plan: Check how keeping info safe works when using learning apps, with rules in mind.
- Goal: To make ways to keep user info safe while keeping the app easy and useful.

### 9. What People Think About Learning on Apps

- Plan: Ask different groups, like students, teachers, heads of schools, and parents, about how they see and hope for learning apps.
- Goal: To understand different views on app learning and find ways to make it better that help everyone.

Limits include just looking at three schools, which might not apply everywhere; using info people said themselves, which may not be true; and a short check time, missing long-term effects. We should look into these limits later. There are many things to check out for this school app. By looking into these things, people who study this can keep helping the app grow, making sure it works well for all types of students and schools. This work will not only make the app better but also teach us a lot about how tech changes learning.

## 10.0. CONCLUSION

Making a new mobile app for higher education marks a big step in meeting the changing needs of students, teachers, and schools in a fast-changing study world. This study shows the big problems in higher education, like how easy it is to get to, how much people engage, how personal it is, and how it helps with jobs, all issues this new app wants to fix well.

By using new tech like AI, data know-how, and tools that make learning feel real, the app makes a well-rounded and changing study zone. It makes getting to study stuff easier and up the level of how much students take part, by using fun interactive parts and talking in real time. Making sure each student has stuff matched to their own way of learning and how they are doing keeps the study path right for them, making a better-fit learning trip.

Also, the app looks after key safety and private info worries, keeping student info safe from people who should not see it and online dangers. Being strong on safety helps build trust with users and gets more people to use the app.

This study thinks that good stuff from the study will help not just with school marks; it helps all of society by making learning open to more people, closing tech gaps, and readying students for the needs of ever-changing work kinds. This app helps keep learning going and builds work skills, making a smarter and well-study group that leads to better money growth and social shifts.

In the end, this study talks about how much mobile apps can change higher education. As schools keep up with new tech, this planned app stands as a key way to better the study times, help people work together, and keep learning through life. More study will be key in making the app better and checking its effects, making sure it stays right for users and fits with today's study goals..

## REFERENCES


Al-Emran, M. (2023). Mobile apps support blended learning adoption in Middle Eastern higher education. *Education and Information Technologies*, *28*(4), 4215–4232. DOI: 10.1007/s10639-022-11431-4

Bates, T. (2015). *Teaching in a digital age: Guidelines for designing teaching and learning*. Tony Bates Associates Ltd.

Braun, V., & Clarke, V. (2021). One size fits all? What counts as quality practice in (reflexive) thematic analysis? *Qualitative Research in Psychology*, *18*(3), 328–352. DOI: 10.1080/14780887.2020.1769238

Campbell University. (2025, March 6). *AI in higher education: A summary of recent surveys of students and faculty*. Academic Technology Blog. https://sites.campbell.edu/academictechnology/2025/03/06/ai-in-higher-education-a-summary-of-recent-surveys-of-students-and-faculty/

Creswell, J. W., & Plano Clark, V. L. (2018). *Designing and conducting mixed methods research* (3rd ed.). SAGE.

Crompton, H. (2013). A historical overview of mobile learning: Toward learner-centered education. In Berge, Z. L., & Muilenburg, L. Y. (Eds.), *Handbook of mobile learning* (pp. 3–14). Routledge.

Kaliisa, R., & Picard, M. (2024). The promise of mobile learning for higher education in Sub-Saharan Africa. *International Journal of Educational Technology in Higher Education*, *21*(2), 45–62. DOI: 10.1186/s41239-024-00456-7

Luckin, R., Holmes, W., Griffiths, M., & Forcier, L. B. (2016). *Intelligence unleashed: An argument for AI in education*. Pearson.

Nguyen, T., Lee, J., & Chen, Y. (2025). Equity gaps in mobile learning adoption: Evidence from an Asian multi-site study. *Computers & Education*, *221*, 105122. DOI: 10.1016/j.compedu.2025.105122

Pachler, N., Bachmair, B., & Cook, J. (2010). *Mobile learning: Structures, agency, practices*. Springer. DOI: 10.1007/978-1-4419-0585-7

Price, D., & Winchester, M. (2025). Blended learning equity in higher education: A scoping review. *Australian Educational Researcher*. Advance online publication. DOI: 10.1007/s13384-025-00859-9

Puentedura, R. R. (2006). Transformation, technology, and education. *Hippasus Blog*. https://hippasus.com/resources/tte/

Raftr (2025). *Mobile-first student engagement in higher education: From apps to AI-powered platforms*. https://www.raftr.com/mobile-first-student-engagement-in-higher-education/

Sajja, R., Reddy, P. K., & Kumar, A. (2023). Artificial intelligence-enabled intelligent assistant for personalized and adaptive learning in higher education. *arXiv preprint*. https://arxiv.org/abs/2309.10892

Siemens, G. (2005). Connectivism: A learning theory for the digital age. *International Journal of Instructional Technology and Distance Learning*, *2*(1), 3–10.

Suresh Babu, C. V., & Adhithya, S. (2023). ChatGPT and Other Generative AI Tools in Education: Transformative Potential and Ethical Considerations. In Keengwe, J. (Ed.), *Creative AI Tools and Ethical Implications in Teaching and Learning* (pp. 135–152). IGI Global., DOI: 10.4018/979-8-3693-0205-7.ch007

Suresh Babu, C. V., & Auroshaa, A. (2024).,Influence of Data Analysis Tools in Teaching and Learning Economics. In *Innovative Pedagogy in the Digital Era: Empowering Learning Through Digital Leadership and Technology Integration* (pp. xx–xx). IGI Global

Suresh Babu, C. V., & Padma, R. (2023). Technology Transformation Through Skilled Teachers in Teaching Accountancy. In González-Lezcano, R. (Ed.), *Advancing STEM Education and Innovation in a Time of Distance Learning* (pp. 211–233). IGI Global., DOI: 10.4018/978-1-6684-5053-6.ch011

Suresh Babu, C. V., & Rohan, B. (2023). Evaluation and Quality Assurance for Rapid E-Learning and Development of Digital Learning Resources. In Santally, M., Rajabalee, Y., & Rajputh, R. (Eds.), *Implementing Rapid E-Learning Through Interactive Materials Development* (pp. 139–170). IGI Global., DOI: 10.4018/978-1-6684-4940-0.ch008

Yang, C., & Taele, P. (2025). Audemy: AI for accessible education for blind students. *arXiv preprint*. https://arxiv.org/abs/2504.17117

Zawacki-Richter, O., Marín, V. I., Bond, M., & Gouverneur, F. (2019). Systematic review of research on artificial intelligence applications in higher education: Where are the educators? *International Journal of Educational Technology in Higher Education*, *16*(1), 39. DOI: 10.1186/s41239-019-0171-0

Zou, Y., Kuek, F., Feng, W., & Cheng, X. (2025). Digital learning in the 21st century: Trends, challenges, and innovations in technology integration. *Frontiers in Education*, *10*, 1562391. DOI: 10.3389/feduc.2025.1562391

# Chapter 4
# Game-Based Learning and Gamification

**Tanya Shakya**
*Institute of Management Studies, Ghaziabad, India*

**Nikita Joshi**
https://orcid.org/0000-0002-1752-2425
*Institute of Management Studies, Ghaziabad, India*

## ABSTRACT

*The integration of game-based learning and gamification into educational environments has revolutionized the way students interact with complex concepts, ushering in an era of dynamic and immersive learning experiences. This chapter invites readers to explore the exciting potential of these innovative methodologies to enhance educational outcomes, particularly in fields that demand practical application and critical thinking, such as cybersecurity and technology. Game-based learning transforms traditional education by turning the learning process itself into an engaging game. In this approach, carefully crafted games are designed to deliver educational content, tapping into the innate motivation that game mechanics inspire. Students are encouraged to dive into interactive worlds where they can explore, experiment, and learn through hands-on experiences and trial-and-error scenarios. Game-based learning is a fun and effective way to keep students engaged while helping them remember what they learn. By weaving together exciting stories and challenging gameplay,*

DOI: 10.4018/979-8-3693-8287-5.ch004

## 1. INTRODUCTION

Game-based learning (GBL) is a dynamic and exciting educational strategy that transforms the way we acquire knowledge by integrating games into the learning process. Imagine immersing yourself in a world where education feels like play, where every challenge and quest is designed to boost your skills and understanding. GBL fosters an environment of active participation and interaction, making learning not just effective, but fun. Gamification is an exciting approach that breathes new life into everyday tasks by infusing them with the engaging elements typically found in games. Imagine turning mundane activities into thrilling challenges where you can earn points, collect badges, climb leaderboards, and embark on rewarding journeys! By harnessing dynamic game mechanics, gamification transforms the ordinary into the extraordinary, making participation more enjoyable and motivating. The essence of gamification lies in our understanding of fundamental human motivations—our natural inclination towards competition, achievement, and social connectivity. This approach finds value across various domains, including education, marketing, employee training, and customer engagement, as it encourages individuals to engage in experiences that are both interactive and rewarding. This chapter invites readers to explore the exciting potential of these innovative methodologies to enhance educational outcomes, particularly in fields that demand practical application and critical thinking, such as cybersecurity and technology (Chandran et al., 2025), (Mhammed, 2025), (Mousa & Shehab, 2025).

It is important to note that while gamification shares some characteristics with game-based learning, it does not involve engaging in complete games. Rather, gamification enhances existing systems by integrating game-like elements, thereby increasing engagement and fostering a passion for learning, mastery, and social interaction. We invite you to explore the transformative potential of gamification in enhancing our interactions with the world, as it can turn even the most routine tasks into captivating experiences. With elements like friendly competition, enticing rewards, and instant feedback, it keeps learners motivated and engaged, enhancing their ability to remember and apply what they've learned (Al-Shareeda et al., 2025; Abdulameer Nayyef Aldabbagh & Ibrahim Zghair Alnasrawi,

2025; Ho et al., 2026). Whether it's solving puzzles, competing in teams, or embarking on epic adventures, GBL turns the classroom into an exhilarating playground for the mind, where every player is a key participant in their educational journey. The transformative power of Game-Based Learning and Gamification lies in the fact that learning feels like an adventure, where every lesson is infused with the excitement of a game. Game-Based Learning (GBL) and gamification are reshaping education by engaging learners through interactive gameplay and inspiring elements that create an immersive experience. These innovative methods not only enhance motivation but also drive deeper understanding of complex concepts. In the realm of GBL, engagement and motivation are the twin engines that propel student learning. The carefully crafted blend of storytelling, challenges, and interactive gameplay acts as a catalyst, igniting students' passion for learning (Alshinwan et al., 2025; Audah Kareem, & Al-Shareeda, 2026a; Audah Kareem, & Al-Shareeda, 2026b). As they embark on quests, strategize their next moves, or race against time to complete missions, students experience a thrill that transcends traditional learning methods. The desire to progress through levels, unlock achievements, and conquer challenges fosters an intrinsic motivation that drives them to delve deeper into the educational content. This active participation transforms the learning process from a chore into an adventure, instilling a lifelong love for exploration and knowledge. This chapter aims to enhance educational outcomes, particularly in fields that demand practical application and critical thinking, such as cybersecurity and technology (Ho et al., 2026; Jaafar et al., 2026; Al-Shareeda & Hawi, 2026; Ramesh et al., 2026; Al-Na'amneh et al., 2025).

## 2. GAME-BASED LEARNING APPROACHES

With GBL, students can confront complex systems and challenges in a safe environment, transforming education into an exhilarating journey. Why settle for one form when you can enjoy both digital and physical experiences? Hybrid games merge tactile and digital elements, captivating various types of learners [Sanchez]. Whether tapping on a screen or interacting with physical components, the thrill of discovery makes learning a dynamic experience. Turn review sessions into epic quests. Engaging games designed to solidify knowledge transform retention into a joyful adventure, complete with challenges that reward success. The playful nature of these games makes the learning journey not only informative but also immensely enjoyable.

### 2.1.1. Role-Playing Games: A New Perspective

Step into someone else's shoes with role-playing games, which are powerful tools for fostering social and emotional growth. These games encourage collaboration, allowing students to tackle moral dilemmas and build empathy. Teamwork takes center stage as participants learn to appreciate perspectives beyond their own, enriching the educational experience.

### 2.1.2.Dynamic Interaction and Immediate Feedback

Imagine a classroom teeming with energy, where students experiment with concepts in real time, gaining instant feedback on their decisions. This vital feedback loop not only enables learners to understand their mistakes but also allows them to celebrate their successes. It's a thrilling way to reinforce knowledge while encouraging repeated exploration.

### 2.1.3.Thrill of Competition and Narrative

Add a sprinkle of competition, and watch the excitement soar! Incorporating leaderboards and peer challenges inspires students to push their limits. This healthy rivalry transforms learning into an exhilarating race for excellence, making every effort feel like part of a rewarding game. But wait, there's more! Engaging narratives breathe life into lessons, turning dull topics into unforgettable adventures. Who wouldn't want to solve a math problem when it's woven into an epic storyline? Educational games with captivating narratives make every lesson feel relatable and enjoyable.

By embracing these innovative methodologies, we can ignite passion and enthusiasm for learning, transforming the educational landscape into an exhilarating playground for knowledge. Experience the thrilling future of education, where fun and learning go hand in hand!

## 3. KEY COMPONENTS OF THE GAMIFICATION DESIGN PROCESS

### 3.1. Define Learning Objectives

The initial phase of the gamification design process entails meticulously defining the desired outcomes for learners. This requires a comprehensive analysis to pinpoint the specific knowledge and skills that are vital for achieving these outcomes. By establishing clear and measurable objectives, designers can tailor gamification elements to align effectively with these goals, ensuring that every game feature contributes to the educational journey (Top Hat, 2019).

### 3.2. Identify Target Audience

Gaining a deep understanding of the target audience is paramount in the design process. This involves researching the demographics, preferences, motivations, and interests of the learners. By employing surveys, interviews, or focus groups, designers can gather insights that inform the creation of gamification elements that not only resonate with the audience but also keep them deeply engaged throughout the learning experience.

### 3.3. Choose Game Mechanics

The selection of appropriate game mechanics is a critical aspect of designing an engaging learning experience. Common mechanics include points systems, achievement badges, leveling up, and various challenges or quests. Each of these elements should be thoughtfully chosen to enhance the overall experience while reinforcing the learning objectives. For instance, incorporating a point system can motivate learners to strive for higher scores, while badges can serve as tangible acknowledgments of their achievements, thereby boosting their confidence and engagement.

## 3.4. Design Game Elements

This phase involves the creation of immersive and interactive elements that elevate the learner's experience. Designers should craft captivating graphics, engaging audio, intricate storylines, relatable characters, and effective feedback systems. These components work in harmony to build a compelling narrative that not only draws learners in but also encourages them to navigate through challenges and achieve their learning goals. The integration of such elements can transform a standard educational activity into an enjoyable adventure.

## 3.5. Test and Evaluate

The concluding phase of the gamification design process is dedicated to testing the newly developed gamified experience. This includes gathering feedback from actual users to assess their experiences and satisfaction levels. It's essential to analyze this feedback critically to identify strengths and areas for improvement. By conducting evaluations and making necessary adjustments, designers can refine the gamified approach for future iterations, ensuring that it remains effective and relevant to the learners' needs.

Gamification and game-based learning (GBL) are two innovative approaches in education designed to enhance student engagement and motivation.

Gamification involves incorporating game-like elements—such as points, badges, leaderboards, and challenges—into the learning process. This approach makes educational activities more appealing, transforming traditional lessons into engaging experiences. By associating learning with excitement and curiosity, gamification encourages students to become active participants in their educational journey. Instead of viewing learning as a chore laden with stress, students begin to see it as a fun and enjoyable process. Game-based learning, on the other hand, uses actual games to create immersive learning experiences. It allows students to engage with educational content in a hands-on way, fostering critical thinking and collaborative skills. Through GBL, students can work together to solve problems, grasp complex concepts, and apply their knowledge in real-world scenarios, all while maintaining a sense of fun. This method not only strengthens motivation but also helps alleviate stress associated with traditional learning environments. Both strategies aim to create a more dynamic and enjoyable classroom atmosphere, ultimately leading to deeper understanding and retention of material (Manuel, 2023) They shift the focus from rote memorization to active participation, helping students develop a love for learning that can last a lifetime. Gamification and game-based learning (GBL) are exciting educational strategies that embrace the world of gaming to elevate learning experiences, but they approach this in unique ways.

## 4. EXCITING EXAMPLES OF GAME-BASED LEARNING

### 4.1 Virtual City Management Simulations

Imagine stepping into the shoes of a city planner, where every decision shapes the destiny of a bustling metropolis. In this immersive simulation game, students become the architects of their own virtual cities. They must think critically about resource allocation, budgeting, and sustainable practices to ensure their city thrives. This hands-on approach not only cultivates an understanding of economic principles but also instills a sense of responsibility toward environmental preservation.

### 4.2 Engaging Educational Puzzles

What if solving math problems was just as thrilling as completing a treasure hunt? In these interactive puzzle games, students embark on an adventure where mathematics is the key to unlocking new levels.[4] By tackling challenges that require problem-solving and logic, learners discover mathematical concepts organically, turning numbers into a thrilling quest rather than a mundane task.

## 4.3 Historical Adventure Games

Imagine stepping into the shoes of a hero navigating the bustling streets of ancient Rome or solving riddles in the majestic temples of Egypt. In these thrilling games, students are not just passive observers; they actively engage with history. As they unravel historical puzzles, they might discover the significance of the Roman Senate or the architectural marvels of the Pyramids. Each puzzle resolved serves as a portal to deeper historical insights, creating a rich tapestry of learning interwoven with adventure. This interactive approach allows students to experience history first-hand, making it not only educational but also incredibly fun.

## 4.4 Science Simulation Games

Picture a virtual laboratory where students play the role of a scientist, tasked with managing a vibrant ecosystem or conducting groundbreaking experiments. In these simulations, learners can manipulate variables, watch ecosystems thrive or collapse, and understand the delicate balance of nature. For instance, they might introduce a new species to an environment and witness its impact, learning about ecological principles in real time. These hands-on experiences demystify complex scientific concepts, making them accessible and engaging, while encouraging students to think critically and creatively as they experiment with different scenarios.

## 4.5 Corporate Training

Gamification has emerged as a transformative strategy in corporate training, helping organizations engage employees and enhance learning outcomes. Companies like Deloitte and AstraZeneca have pioneered this approach with impressive results. For instance, Deloitte's Leadership Academy employs a gamified model that features leaderboards, badges, and points systems to encourage participation and healthy competition among employees. This not only increases engagement but also fosters a culture of achievement and motivation.

AstraZeneca took a different route with their Go-to-Jupiter system, which gamified the training experience by incorporating levels and rewards tied to the completion of training modules. This method significantly boosted the completion rates for essential trainings, allowing employees to feel a sense of progression and accomplishment as they advance through various levels of their learning journey. By integrating elements typically found in games, these companies have turned training into an interactive experience that keeps employees motivated and eager to learn.

## 4.6 Classrooms

In educational settings, gamification has proven to be a powerful tool in enhancing student engagement and learning retention. Platforms such as Duolingo have revolutionized language learning by introducing game-like elements into their structure. By setting daily goals, employing leaderboards, and providing badges for achievements, Duolingo brings a competitive yet supportive atmosphere to language acquisition. This not only makes learning more enjoyable but also encourages users to practice regularly and strive for continuous improvement in their language skills.

Teachers are also embracing gamification in their classrooms by incorporating game mechanics into lesson plans. For example, educators might use point systems for assignments, create quizzes that simulate game environments, or offer rewards for academic milestones. These strategies can increase student participation and foster a sense of community within the classroom, transforming learning into a more dynamic and engaging experience.

## 4.7 Online Courses

The rise of eLearning has opened up new avenues for gamification. Many online course providers are enhancing their platforms by integrating quizzes, interactive challenges, and progress tracking systems that reward learners for their efforts. Through the use of points and badges, learners can visualize their progress and stay motivated as they tackle complex subjects.

- Assassin's Creed: While not primarily designed as an educational tool, the Assassin's Creed series immerses players in various historical settings, offering a unique opportunity to learn about significant events and cultures. The game's detailed environments, characterized by accurate historical landmarks and events, can serve as a supplementary resource for teaching about different eras, including the pivotal moments of the Middle Ages. Educators can use the game to spark discussions about the historical context of events, societal structures, and cultural art forms of the time, making history come alive for students.
- Europa Universalis: This grand strategy game focuses on detailed simulations of historical geopolitics, allowing players to assume the role of a nation and navigate through centuries of history. Europa Universalis promotes critical thinking and decision-making skills as players engage with complex historical scenarios. Students can explore themes such as diplomacy, war, trade, and colonization while learning about the interplay of factors that shaped world history. This game provides an interactive platform where players develop an understanding of historical cause and effect, the impact of leadership decisions, and the complexities of international relations.

## 4.8 Pentiment

Set against the backdrop of early 16th-century Germany, Pentiment offers an immersive narrative-driven experience that revolves around a richly detailed art style influenced by medieval manuscripts and early prints. Players engage in mystery-solving and historical investigation, which invites them to delve deeper into the social, political, and religious dynamics of the period. The game serves as a tool for introducing students to key themes such as the Reformation, the impact of printing on society, and the evolution of artistic expression. This exploration encourages players to think critically about historical narratives and the sources that shape our understanding of the past.

For instance, some platforms use gamified assessments that allow learners to earn points for completing modules or participating in discussions, which can be redeemed for additional resources or discounts on future courses. Interactive challenges and simulations can also make tough topics more approachable, turning potentially overwhelming material into enjoyable learning experiences. By fostering a sense of achievement and providing immediate feedback, gamification in online learning environments helps to maintain student motivation and encourages a deeper understanding of the material. While gamification sprinkles game-like elements into traditional educational settings—like rewards or competitions—Game-Based Learning (GBL) dives headfirst into an exciting world where education is woven into every aspect of the gameplay. In GBL, learning isn't just a side quest; it's the main mission! Gamification harnesses the power of external motivators, like points and badges, to inspire students. Conversely, GBL ignites intrinsic motivation, captivating learners through a sense of curiosity and the thrill of exploration. It's all about making learning feel like an adventure rather than a chore. Gamification tends to reward players with tangible incentives, creating friendly competition, while GBL fosters an environment driven by wonder and discovery. Students become explorers on a quest, their desire to learn fueled by their innate curiosity rather than external accolades. Through these dynamic frameworks, education transforms from a passive experience into an exhilarating journey, making learning more engaging and effective than ever before. Gamification and game-based learning (GBL) are transforming education into an exhilarating adventure. Imagine stepping into a classroom where lessons feel more like quests and achievements unlock rewards, captivating students like never before. The magic of gamification lies in its ability to turn mundane tasks into thrilling challenges, sparking a zest for learning that keeps students coming back for more.

GBL elevates this excitement to a whole new level, immersing learners in fascinating worlds where they can explore, discover, and conquer new knowledge at their own pace. Imagine customizing difficulty levels to match each student's unique abilities—this level of personalization empowers learners and fosters a deeper connection with the material. Together, gamification and GBL are not just educational strategies; they are catalysts for change, reimagining how we engage with information. By blending the allure of games with scholarly pursuits, we're creating an environment where learning isn't just effective; it's a thrilling journey. This innovative approach has the potential to revolutionize education, making it not only more engaging but also profoundly impactful. Overall, gamification is reshaping how we approach training and education across various contexts, making learning more engaging, competitive, and rewarding. It leverages our innate desire for achievement, recognition, and interaction, paving the way for more effective and enjoyable learning experiences

## 4.9 Real-World Examples of Gamification in Action

- Classroom Challenges: Think of assignments as quests where students earn points and unlock special privileges, creating a sense of adventure in their learning journey.
- Virtual Scavenger Hunts: Students embark on exciting online expeditions to hunt down educational treasures, racing against the clock to earn points and badges — a thrilling way to absorb knowledge!

## 4.10 Game-Based Learning (GBL): Learning Through Play

In Game-Based Learning, where games aren't just a tool — they are the very heart of the learning experience GBL invites students into immersive gameplay, requiring them to solve problems and apply skills within the game's universe. Whether it's navigating simulations, cracking challenging puzzles, or unraveling interactive stories, GBL transforms learning into an experiential adventure. Students aren't just passive recipients of information; they become active explorers, diving deep into complex concepts and discovering knowledge through hands-on experiences. In this vibrant atmosphere, every quest, every challenge becomes an opportunity for growth, allowing students to thrive in an engaging and interactive environment. So, whether it's gamifying the classroom experience or jumping into learning games, the possibilities are endless harness the power of play to ignite a passion for learning and create unforgettable educational adventures!

## 5. ENVIRONMENT FOR GAME-BASED LEARNING

In the classroom, Gamification enhances existing environments, breathing new life into traditional educational settings. It doesn't require a complete overhaul; instead, it spices up the classroom by weaving in elements that make the learning experience more engaging and enjoyable.

Points, badges, leaderboards, and challenges—these game-like features are not just fun additions; they're powerful motivators. By integrating these elements into everyday activities, students find themselves more inspired and eager to participate, turning lessons into thrilling quests rather than mundane tasks. Gamification respects the traditional structure of education. It keeps essential learning objectives and curriculum in place, ensuring that while the excitement ramps up, students stay on track to meet educational standards. It's like putting a turbo boost on learning—where engagement and achievement go hand in hand!

## 6. FUNCTIONING OF GAMIFICATION

Gamification thrives on familiarity. By building upon structures students already know, such as levels and rewards, it creates a cozy learning environment. This sense of comfort allows them to immerse themselves in the content, rather than feeling lost in a new and confusing system. With features like leaderboards and badges, gamification provides instant feedback that keeps students on their toes. They can track their own achievements and compare their progress with their peers, igniting a friendly competition that fuels further engagement. Imagine the excitement of striving to climb the leaderboard or earning that shiny badge—learning becomes not just informative, but exhilarating. In essence, gamification transforms education into a captivating experience where every student can thrive and have fun while learning. Gamification is transforming the way we engage with learning, and one standout example is Kahoot This dynamic platform energizes the classroom by allowing educators to design interactive quizzes and games that feel more like fun competitions than traditional assessments. Imagine students eagerly participating in live sessions, their hearts racing as they race against their classmates to answer questions correctly. With bright visuals, catchy music, and the thrill of scoring points, Kahoot! makes learning exciting and memorable. It's not just about getting the answers right—it's about the adrenaline rush of friendly rivalry that drives student engagement and deepens understanding in a way that textbooks simply can't match (Top Hat, 2019).

Also, assignments are transformed into exciting challenges, allowing students to earn points or badges for completing various tasks. These accumulated points can be used to unlock unique privileges or rewards, enhancing engagement and motivation in the learning process. This approach not only makes learning more enjoyable but also encourages students to actively participate and excel in their educational journey.

## 6.1 Key Aspects of Game-Based Learning (GBL)

### a) Immersive Game Worlds

At the heart of Game-Based Learning (GBL) lies the creation of immersive game worlds that whisk students away from the confines of their traditional classrooms and into rich, interactive environments. Imagine stepping into a vibrant virtual world where learners can traverse bustling marketplaces of ancient civilizations, roam expansive fantasy realms, or explore distant planets—experiences that make the learning journey feel like an exciting adventure. These environments are meticulously crafted to captivate the imagination, filled with stunning visuals and ambient sounds that draw students in, making them feel like active participants in the narrative rather than passive observers. The allure of these game worlds lies in their ability to create a sense of agency; students aren't just learning—they are explorers, problem-solvers, and heroes in their own educational stories.

### b) Rich Narratives

The narratives embedded within these games are far from ordinary; they are intricately woven tales that resonate with the learners' aspirations and curiosity. Through compelling storylines, students may find themselves thrust into critical moments of history, tasked with making decisions that can alter the course of an ancient civilization or unveiling the mysteries of science as they engage with complex characters and scenarios. For instance, imagine a game where learners must collaborate as a team of historians, piecing together clues to solve a historical enigma or navigate ethical dilemmas in a futuristic society. These narratives not only enhance engagement but also provide context for the learning objectives, immersing students in experiences that are relevant and meaningful.

## c) Enticing Challenges

Embedded within these immersive narratives are a plethora of enticing challenges that require ingenuity, collaboration, and critical thinking. As students navigate through intricate puzzles, they are not just playing a game; they are applying the knowledge and skills they have acquired in creative ways. Each challenge is thoughtfully designed to inspire curiosity and push the boundaries of what students believe they can achieve. Picture a scenario where students must decode ancient scripts to unlock the next level of their quest, all while drawing upon their understanding of linguistics or cultural history (Pan et al., 2021). These challenges are inherently rewarding, encouraging persistence and a growth mindset as students experience the thrill of overcoming obstacles and celebrating their discoveries. The feedback provided throughout these challenges adds another layer of depth, offering insights and encouragement that motivate learners to explore further.Game-Based Learning is revolutionizing education by unlocking the exciting potential of immersive games, captivating stories, and thrilling challenges. This innovative approach not only sparks enthusiasm but also cultivates a lifelong passion for learning that goes beyond traditional classroom walls. It equips students with the curiosity and creativity needed for a future rich in exploration and innovation.

## d) Interactive Learning: The Heart of Engagement

Imagine stepping into a vibrant game world where you're not just a spectator, but an active explorer in your own educational adventure. At the heart of Game-Based Learning lies interactive learning, which transforms the way students engage with knowledge. Instead of merely absorbing facts, learners dive headfirst into an exhilarating game environment filled with tailored challenges that masterfully blend fun with skill development.

Whether you're piecing together intricate puzzles, charting courses through mysterious realms, or teaming up with classmates to conquer obstacles, every interaction is crafted to ignite critical thinking and enhance problem-solving abilities. This hands-on experience not only heightens cognitive engagement but also encourages students to experiment with different strategies and viewpoints as they traverse the game's many landscapes. Get ready to embark on an educational journey where learning is as thrilling as the game itself.

### e) Immersive Experience

Where Emotion Meets Intellect: At the heart of Game-Based Learning (GBL) lies an extraordinary ability to transport students into vibrant, immersive worlds that captivate both their hearts and minds. Imagine stepping into the shoes of heroic figures in a historical simulation, navigating the complexities of their challenges and victories. As players embark on these thrilling quests, they find themselves not just observing history but living it, forging an emotional bond with the struggles and triumphs that shaped our world (Adipat et al., 2021).

This exhilarating journey transforms learning into an adventure, transforming abstract concepts into tangible experiences. The moment a student faces the moral dilemmas of key historical figures or collaborates with peers to overcome obstacles, they are more than just participants; they become part of the narrative. Such profound engagement not only heightens retention but also empowers learners to draw fascinating parallels between the game's challenges and real-life scenarios, enriching their educational experience. In a realm where imagination knows no bounds, Game-Based Learning invites students to explore, discover, and connect, making education not just a pursuit of knowledge but an unforgettable journey of personal growth and transformation.

# 7 BENEFITS OF GAME-BASED LEARNING

## 7.1 Increased Engagement

GBL transforms traditional learning into an exciting adventure, captivating students' interest in ways that lectures and textbooks often can't. With enticing storylines and interactive components, students look forward to learning, making the educational process more dynamic and enjoyable.

## 7.2 Improved Retention

The immersive experiences provided by GBL create lasting memories. When students engage in active problem-solving and exploration, they are more likely to retain the knowledge gained. The interactive nature of these games helps solidify concepts in a way that mere memorization cannot match. Imagine students effortlessly recalling historical events or scientific facts because they've experienced them in a vivid, game-like setting.

## 7.3 Personalized Learning

One of the most powerful aspects of GBL is its adaptability. Many platforms can adjust the level of difficulty based on individual student performance, ensuring that everyone is challenged yet supported. This personalized approach allows for differentiated learning paths, where students can progress at their own pace, building confidence as they master each level before moving on to more complex challenges.

## 7.4 Game-Based Learning is Revolutionizing Education by Transforming the Way Students Engage With Complex Concepts

It turns classrooms into immersive worlds filled with rich narratives, challenges, and opportunities for exploration. With GBL, learning becomes an adventure, where students are not merely passive recipients of information but active participants in their educational journeys. By engaging their minds and hearts, GBL captivates students and inspires a lifelong love of learning. Whether they are navigating the trials of history or experimenting in a virtual lab, students discover the joy of knowledge, preparing them for a future that is both informed and imaginative.[Adipat.]

## 8 IMPLEMENTING GAMIFICATION EFFECTIVELY IN LEARNING CONTEXTS

Gamification, the application of game design elements in non-game contexts, has gained significant traction in educational settings as a way to engage learners and enhance motivation. To effectively integrate gamification into learning experiences, consider the following essential strategies:

### 8.1. Define Clear Objectives

Establishing explicit goals is critical when implementing gamification. These objectives should align closely with the learning outcomes you wish to achieve. For instance, if the aim is to improve problem-solving skills, gamified activities should focus on scenarios that challenge learners to apply their knowledge in practical contexts. Clear objectives help ensure that gamified elements are not merely engaging but also serve to reinforce the educational purpose, providing a structured path for learners to absorb and apply new concepts.

### 8.2. Understand Your Audience

Knowing your audience is paramount. Different learners come with varied backgrounds, experiences, and preferences. For effective gamification, it's important to consider diverse learning styles—visual, auditory, kinesthetic—and develop experiences that cater to these differences. Conduct surveys or focus groups to gather insights on what motivates your learners and which game elements they find appealing (for instance, rewards, challenges, or narratives). Creating personas based on this data can help tailor the gamified experience, ensuring it resonates with all participants and enhances their engagement.

## 8.3. Use Existing Platforms

To streamline the integration of gamification, leverage existing training platforms rather than starting from the ground up. Many modern e-learning platforms already offer gamified features such as badges, leaderboards, and progress tracking. Utilizing these tools can save time and resources while ensuring that the gamification elements are seamlessly aligned with the curriculum. Additionally, this approach can facilitate an immediate sense of community and competition, further encouraging participation and collaboration among learners.

## 8.4. Incorporate Feedback Mechanisms

An effective gamification strategy includes a system for providing learners with regular feedback. This can take the form of points earned, levels achieved, or constructive critiques on performance. Feedback not only reinforces learning but also helps learners understand their progress and areas that need improvement. Providing timely and constructive feedback keeps learners motivated and allows them to adapt their strategies accordingly.

## 8.5. Foster a Collaborative Environment

While competition can drive engagement, collaboration is equally important in enhancing the learning experience. Incorporating team-based gamification activities encourages communication, teamwork, and peer-to-peer learning. For instance, learners can work together on problem-solving challenges, share strategies, and celebrate team victories. This not only makes the learning process more enjoyable but also builds essential soft skills that are valuable in real-world scenarios.

## 8.6. Iterate and Improve

Finally, gamification is not a one-size-fits-all solution. To ensure continued effectiveness, regularly assess and refine your gamified elements based on feedback from participants and performance analytics. Encourage learners to share their experiences and suggestions for improvement. Flexibility and willingness to adapt can greatly enhance the success of gamification initiatives, making the learning environment more responsive to learner needs (Hartt et al., 2020).

By embracing and thoughtfully implementing these strategies, educators and trainers can create dynamic learning environments that inspire participation, foster growth, and transform learning into an enjoyable adventure. Gamification, when executed effectively, not only enhances engagement but also cultivates a culture of continuous learning that prepares individuals for future challenges.

## 9. KEY ELEMENTS OF HOLISTIC GBL DESIGN

### 9.1. Integration of Educational Objectives

At the core of GBL is the integration of clear educational outcomes directly into the gameplay mechanics. Unlike traditional educational methods, where learning may sometimes feel disconnected from practice, GBL ensures that learners actively apply knowledge as they play. For example, a game centered around environmental science might require players to solve ecological problems using principles of biology and chemistry. This interactive method promotes retention and understanding as students are not merely memorizing facts; they are using knowledge in a relevant context.

### 9.2. Personalization and Adaptability

One of the remarkable advancements in GBL platforms is their ability to adapt to individual learners' needs. By leveraging data analytics and AI, these games can assess a player's strengths, weaknesses, and learning pace in real-time. If a student struggles with a particular concept, the game can dynamically modify challenges to provide additional practice or alter the complexity of the tasks. This kind of personalized learning experience increases both efficacy and engagement, as it accommodates diverse learning styles and helps students build confidence at their own pace.

## 9.3. Promotion of Critical Thinking and Problem-Solving

GBL transforms learners into active problem solvers by presenting them with scenarios that require careful analysis and decision-making. For instance, puzzle-based games can simulate real-world challenges, encouraging students to think critically and employ strategies to navigate obstacles. This not only develops cognitive skills but also prepares students for future endeavors, fostering resilience and adaptability. Such skills are invaluable in both academic and everyday contexts, where problem-solving and critical thinking are frequently required.

## 9.4. Enhanced Engagement and Motivation

One of the standout features of GBL is its ability to captivate learners through interactive and immersive experiences. Unlike conventional learning methods, which can sometimes feel monotonous or passive, games are designed to be inherently engaging. Elements like rewards, leaderboards, and achievements create a sense of progression and accomplishment. This gamification can lead to increased motivation, as students are more likely to persist through challenges when they can see their advancements recognized and celebrated. Furthermore, the social aspect of multiplayer games can enhance motivation, as learners collaborate and compete with peers, fostering a community of learners (Alfares, 2025)

In essence, Game-Based Learning transforms the educational landscape by crafting experiences that are not only educational but also enjoyable. Through the integration of learning objectives, adaptability to individual needs, promotion of critical thinking skills, and enhancement of motivation, GBL stands out as a powerful method for captivating learners and enriching their educational journey. As technology continues to evolve, the possibilities for GBL are vast, presenting exciting opportunities for educators and students alike to explore new horizons in learning.

## 10. GAMIFICATION FRAMEWORKS

In the realm of gamification design, several established frameworks provide guidance, each with a unique focus that enhances the overall experience:

1. Player-Centred Design: This approach emphasizes a deep understanding of the player's unique context, preferences, and objectives. By aligning the gamification elements with the player's needs, designers can create experiences that resonate personally and promote meaningful engagement.
2. Octalysis Framework: Developed by Yu-kai Chou, this framework breaks down gamification into eight core elements—Epic Meaning, Accomplishment, Empowerment of Creativity and Feedback, Ownership, Social Influence, Scarcity, Unpredictability, and Loss & Avoidance. Each element contributes to crafting experiences that captivate users, facilitating a more immersive and motivating interaction.
3. Self-Determination Theory (SDT): This psychological theory posits that individuals have three innate psychological needs—autonomy, competence, and relatedness—that must be satisfied to foster motivation and engagement (Patterson, 2024). By incorporating these elements into gamified systems, educators can transform participation into a self-driven and enjoyable experience that promotes sustained involvement.

By weaving together these strategies, educators and designers can design engaging learning experiences that captivate learners' interest, transforming traditional study sessions into exhilarating quests for knowledge and achievement.

## 11. BENEFITS OF HOLISTIC GBL DESIGN

1. Durable Learning Outcomes: One of the standout benefits of GBL is its emphasis on meaningful decision-making. When learners are faced with choices that have tangible consequences within the game,

they gain a more profound understanding of the subject matter. This immersive approach instills not just rote memorization but a deeper conceptual grasp, making it easier for students to retain information over time. Studies have indicated that such experiences lead to more sustainable learning compared to traditional educational methods.
2. Development of Lifelong Learning Skills: GBL fosters a growth mindset by encouraging students to embrace challenges and view setbacks as opportunities for learning. Through gameplay, they encounter obstacles that require them to think critically and creatively (Tahir & Wang, 2022). This cultivates resilience—teaching learners that persistence is key to overcoming difficulties. As students navigate their own learning paths in a dynamic environment, they develop autonomy in their educational pursuits, empowering them to take ownership of their learning experiences long into the future.
3. Positive Learning Environment: One of the most significant features of GBL is the safe and supportive environment it creates. In this space, experimentation is encouraged, and mistakes are seen as a natural part of the learning process. This reduces anxiety around failure, enabling students to explore new concepts and ideas freely. In turn, this promotes a culture where innovation and curiosity thrive, making learning more engaging and less intimidating.

# 12. ASSESSMENT TOOLS IN GAME-BASED LEARNING (GBL)

## Exploring Assessment Tools in Game-Based Learning (GBL)

Game-Based Learning (GBL) has transformed the way we view education by merging engaging gameplay with essential educational content. This dynamic approach not only captivates learners but also incorporates innovative assessment tools that are crucial for evaluating student progress and understanding. Let's dive into the fascinating world of assessment tools within GBL and explore how they enhance the learning experience:

## 12.1. Formative Assessment: The Adventure of Discovery

Picture this: as players embark on their educational quests, they encounter challenges that adapt to their skill level. Formative assessments serve as a continual checkpoint in the game, providing instant feedback on their performance. Imagine a player facing a clever riddle that adjusts in complexity based on previous answers, creating a tailored learning experience. This way, learners can explore concepts without the stress of traditional exams, making every step feel like a stepping stone to mastery.

## 12.2. Summative Assessment: The Epic Finale

At the culmination of each gaming module, students face their grand challenge—a summative assessment that tests their accumulated knowledge and skills. This could involve an epic showdown or building a unique project that showcases their abilities. For example, players might need to harness multiple strategies to conquer a tough level, providing educators a window into their overall comprehension while keeping the excitement alive.

## 12.3. Adaptive Learning: Your Personalized Guide

The beauty of GBL lies in its ability to adapt, much like a skilled game master. By employing adaptive learning technologies, games can modify challenges based on each player's journey. When a student finds themselves struggling with a particular concept, the game instinctively steps in with extra resources and tailored tasks to guide them back on track, ensuring no player is left behind.

## 12.4. Analytics and Data Tracking: The Hidden Treasure Map

Behind the scenes, game analytics serve as a treasure map for educators, revealing insights about student engagement and performance in real-time. Imagine having access to dashboards filled with valuable information on completion rates and areas needing improvement—all of which can inform instructional decisions. This data-driven approach promotes a personalized learning journey, making every player's experience unique.

## 12.5. Peer and Self-Assessment: Building a Learning Community

GBL fosters collaboration, encouraging players to engage in peer assessments where they can critique each other's strategies and solutions. This not only enhances critical thinking but also builds a sense of community. Self-assessment tools integrated into games empower students to reflect on their growth and learning, making them active participants in their educational journey.

## 12.6. Game Mechanics as Assessment: Leveling Up Learning

The thrill of gaming lies in its mechanics—like earning points, leveling up, or unlocking achievements. These features double as effective assessment tools, giving players a tangible sense of accomplishment as they master new skills. This intrinsic motivation keeps students hooked and eager to dive deeper into the learning material, transforming assessments into rewarding experiences.

## 12.7.Feedback Loops: Instant Insights

One of the standout features of GBL is its ability to provide immediate feedback, creating a dynamic learning environment. Players quickly see the impact of their decisions, which reinforces learning and cultivates a growth mindset. Mistakes become opportunities for growth, and every failure becomes a chance to strategize anew in a safe and supportive space.

## 12.8. Standardized Assessments: Bridging the Gap

While the playful nature of GBL engages students, incorporating standardized assessments into the mix ensures we can compare results across different contexts. By aligning game content with curriculum standards, educators can verify that learning objectives are being met while still inviting students to enjoy their educational adventures (Mittman et al., 2024).

In essence, assessment tools in Game-Based Learning are the secret ingredients that spice up the educational journey. With a variety of strategies at their disposal, educators can deliver personalized feedback that adapts to each student's needs, ultimately transforming the learning experience into an exciting and engaging adventure. Through GBL, we embark on a mission not just to measure learning but to celebrate it.

One of the key advantages of game-based learning is the use of built-in assessment tools. These tools play a vital role in tracking player progress and engagement. By analyzing player interactions, educators can identify areas where learners excel and pinpoint aspects that may require additional support or resources. This adaptive feedback mechanism allows for a more personalized learning experience, ensuring that each student receives the guidance they need to master the material. It transforms the learning process into a dynamic feedback loop, where educators can tailor their instructional strategies based on real-time data from the games.

# 13. CONCLUSION

In summary, both gamification and game-based learning share the common goal of enhancing engagement and motivation, but they do so through different lenses. Gamification sprinkles enjoyment onto existing learning experiences, while GBL constructs entire realms of adventure and exploration, making learning an exhilarating journey. Game-Based Learning (GBL) make learning enjoyable and also encourage essential social interactions. Students are prompted to work together to solve problems, strategize, and communicate effectively. Such interactions cultivate vital skills including active listening, empathy, leadership, and conflict resolution. These skills are fundamental not only in academic settings but also in building relationships and achieving success in professional environments. By learning to collaborate within a structured framework, students gain confidence in their ability to engage with others, which can have lasting benefits well beyond the classroom.

By leveraging the immersive elements of these games, GBL can cultivate an engaging educational atmosphere that does more than convey historical facts. It invites students to actively participate in their learning journey, fostering curiosity and a deeper appreciation for the complexities of history. The interactive nature of games helps to promote retention of knowledge, as players are likely to remember experiences they actively engage with rather than passively absorb. Furthermore, GBL nurtures collaboration and communication skills as students often work together within the game environments, fostering teamwork and shared problem-solving. This blend of entertainment and education not only enhances students' understanding of specific historical periods, such as the medieval era, but also equips them with critical skills that are valuable in today's world. Overall, game-based learning represents a transformative approach to education, where enjoyment and learning go hand-in-hand.

## 14. IMPLEMENTATION AND FUTURE DIRECTIONS

As educational technology continues to advance, the potential for GBL to evolve is immense. Future GBL initiatives are likely to harness the capabilities of augmented reality (AR) and virtual reality (VR) to create even more immersive learning experiences. These technologies could allow students to engage with complex subjects in a three-dimensional context, simulating real-world scenarios and providing hands-on experiences that are not viable in traditional settings.

Moreover, as research in the field progresses, GBL designs will increasingly be refined to better cater to diverse learners. Customizable learning pathways, adaptive difficulty levels, and culturally relevant content are likely to be integrated to ensure that GBL meets the varied needs of students across different educational stages and backgrounds. This commitment to inclusivity will be crucial in maximizing the effectiveness of GBL as a tool for modern education, ensuring that all learners can benefit from its innovative approach.

In summary, GBL represents a transformative approach to education, one that emphasizes collaboration, personal growth, and an adaptable learning environment—key components in preparing students for the complexities of the real world.

## REFERENCES


Abdulameer Nayyef Aldabbagh, Y., & Ibrahim Zghair Alnasrawi, A. (2025). A Comparative Analysis of Classification Algorithms in Land Cover Mapping: A Study of MLC, Mahalanobis Distance, NNC, and SVM for 2023 and 2018: Case study of Wasit province, central Iraq. *Jordanian Journal of Informatics and Computing*, *2025*(1), 49–66. DOI: 10.63180/jjic.thestap.2025.1.6

Adipat, S., Laksana, K., Busayanon, K., Ausawasowan, A., & Adipat, B. (2021). Engaging students in the learning process with game-based learning: The fundamental concepts. *International Journal of Technology in Education*, *4*(3), 542–552. DOI: 10.46328/ijte.169

Al-Na'amneh, Q., Aljawarneh, M., Alhazaimeh, A. S., Hazaymih, R., & Shah, S. M. (2025). Securing Trust: Rule-Based Defense Against On/Off and Collusion Attacks in Cloud Environments. *STAP Journal of Security Risk Management*, *2025*(1), 85–114. DOI: 10.63180/jsrm.thestap.2025.1.5

Al-Shareeda, M. A., & Hawi, I. A. (2026). IoT-Driven Enterprise Development and Supply Chain Risk Management in Healthcare Organizations. *STAP Journal of Security Risk Management*, *2026*(1), 98–111. DOI: 10.63180/jsrm.thestap.2026.1.7

Al-Shareeda, M. A., Obaid, A. A., & Almajid, A. A. H. (2025). The Role of Artificial Intelligence in Bodybuilding: A Systematic Review of Applications, Challenges, and Future Prospects. *Jordanian Journal of Informatics and Computing*, *2025*(1), 16–26. DOI: 10.63180/jjic.thestap.2025.1.3

Alfares, N. S. (2025). Investigating the efficacy of WordWall platform in enhancing vocabulary learning in Saudi EFL classroom. *International Journal of Game-Based Learning*, *15*(1), 1–12. DOI: 10.4018/IJGBL.367870

Alshinwan, M., Memon, A. G., Ghanem, M. C., & Almaayah, M. (2025). Unsupervised text feature selection approach based on improved Prairie dog algorithm for the text clustering. *Jordanian Journal of Informatics and Computing*, *2025*(1), 27–36. DOI: 10.63180/jjic.thestap.2025.1.4

Audah Kareem, K., & Al-Shareeda, M. A. (2026a). A Low-Complexity Li-Fi Communication Framework for Short-Range Text Transmission. *Jordanian Journal of Informatics and Computing*, *2026*(1), 15–24. DOI: 10.63180/jjic.thestap.2026.1.2

Audah Kareem, K., & Al-Shareeda, M. A. (2026b). A Cost-Effective Embedded Weather Monitoring System for Airport Air Traffic Control Applications. *Jordanian Journal of Informatics and Computing*, *2026*(1), 38–46. DOI: 10.63180/jjic.thestap.2026.1.4

Chandran, A. L., Samual, J., Safavi, S., & Ali, A. (2025). A Comparative Analysis of Machine Learning Models on Detecting Malware in Android Devices. *Journal of Cyber Security and Risk Auditing*, *2025*(4), 327–346. DOI: 10.63180/jcsra.thestap.2025.4.10

Hartt, M., Hosseini, H., & Mostafapour, M. (2020). Game on: Exploring the effectiveness of game-based learning. *Planning Practice and Research*, *35*(5), 589–604. DOI: 10.1080/02697459.2020.1778859

Ho, M., Ang, S., Huy, S., & Janarthanan, M. (2026). Cybersecurity Risks and Challenges in Smart Cities: A Review with Insights for Cambodia. *STAP Journal of Security Risk Management*, *2026*(1), 87–97. DOI: 10.63180/jsrm.thestap.2026.1.6

Jaafar, H. S., Abed, A. A., & Al-Shareeda, M. A. (2026). A Secure Industrial Internet of Things (IIoT) Framework for Real-Time PI Control and Cloud-Integrated Industrial Monitoring. *STAP Journal of Security Risk Management*, *2026*(1), 77–86. DOI: 10.63180/jsrm.thestap.2026.1.5

Manuel. (2023, December 15). 8 game based learning examples. Teach Find. https://teachfind.com/teaching-strategies/game-based-learning/

Mhammed, K. (2025). Cyber Intelligence & Moroccan National Security: What Strategy for Managing and Mitigating Cyber Threats against Future Sporting Events. *Journal of Cyber Security and Risk Auditing*, *2025*(4), 366–383. DOI: 10.63180/jcsra.thestap.2025.4.12

Mittmann, G., Barnard, A., Zehetmayer, S., Wimmer, S., Zehetner, V., Doerfler, S., Woodcock, K., & Schrank, B. (2024). A serious game for emotion regulation in adolescents. *International Journal of Game-Based Learning*, *14*(1), 1–17. DOI: 10.4018/IJGBL.358940

Mousa, R. S., & Shehab, R. (2025). Applying risk analysis for determining threats and countermeasures in workstation domain. *Journal of Cyber Security and Risk Auditing*, *2025*(1), 12–21. DOI: 10.63180/jcsra.thestap.2025.1.2

Pan, L., Tlili, A., Li, J., Jiang, F., Shi, G., Yu, H., & Yang, J. (2021). How to implement game-based learning in a smart classroom? A model based on a systematic literature review and Delphi method. *Frontiers in Psychology*, *12*, 749837. Advance online publication. DOI: 10.3389/fpsyg.2021.749837 PMID: 34925153

Patterson, J. M. (2024). Repurposing video games as discussion tools. *International Journal of Game-Based Learning*, *14*(1), 1–19. DOI: 10.4018/IJGBL.349589

Ramesh, H., Ismail, N., Abd Rahman, N. A., & Ali, A. (2026). PhishGuard: AI-Driven Graph-Based Analysis for Smarter Email Security. *STAP Journal of Security Risk Management*, *2026*(1), 31–45. DOI: 10.63180/jsrm.thestap.2026.1.2

Tahir, R., & Wang, A. I. (2022). Evaluating the effectiveness of game-based learning for teaching refugee children Arabic using the integrated LEAGUÊ-GQM approach. *Behaviour & Information Technology*, *43*(1), 110–138. DOI: 10.1080/0144929X.2022.2156386

Top Hat. (2019, September 16). Game based learning definition and meaning. https://tophat.com/glossary/g/game-based-learning/

# Chapter 5
# Innovative Pedagogies in Online and Blended Learning Environments: Transforming Education in the Digital Age

**Seema Yadav**
https://orcid.org/0000-0002-8477-5895
*Bhopal School of Social Sciences, Bhopal, India*

## ABSTRACT

*The evolving landscape of online and blended learning is being shaped by technological advancements and an increasing emphasis on personalized education, flexibility, and accessibility. The integration of artificial intelligence, AR, and VR is enhancing immersive and adaptive learning experiences This chapter explores the transformative impact of online and blended learning on educational practices, emphasizing the theoretical frameworks, advantages, methodologies, assessment strategies, andfuture trends shaping this dynamic landscape. It highlights the role of constructivist and cognitive theories in designing learner-centered environments that promote engagement and collaboration. It examines the pedagogical innovations, technological advancements, and instructional strategies that are redefining the educational experience. The chapter provides a*

DOI: 10.4018/979-8-3693-8287-5.ch005

*comprehensive analysis of the benefits, challenges, and future directions of online and blended learning, drawing on recent research and case studies to highlight effective practices and emerging trends.*

## INTRODUCTION

The goal of the educational system, which adapts to the advancements in educational technology, is to develop students' ability to think critically and solve problems on their own, as well as to become more actively involved in their education. Critical thinking is a type of thinking that involves the mental processes of judgment, analysis, and assessment. Combining online and in-person learning methods is known as blended learning (Akyüz & Samsa, 2009). Educating in a millennial-dominated world is a difficult undertaking for teachers at Institutes of Higher Learning (IHLs). The development of digital learning, which has enhanced learning techniques in higher education, is one characteristic that makes teaching millennials unique. In many courses, the blended learning technique has been implemented in addition to in-person contact with instructors to meet the needs of millennial learners (Seman et al., 2019). Advancements in technology alter the configuration of educational settings. Web-based learning activities are now transforming learning environments into a mixed structure that combines an online and face-to-face environment. Face-to-face and online settings have long been compared in the context of their instructional capabilities in the field of educational technology (Mese & Dursun, 2019). In Finland nowadays, new educational technology plays a significant role in the training of teachers (Ruokonen & Ruismäki, 2016). The main factors that contributed to the adoption of mobile learning were learning content quality, content design quality, interaction, functionality, user-interface design, accessibility, personalization, and responsiveness. These factors had a favorable impact on students' perceptions of their beliefs (Almaiah et al., 2016a).

All levels of China's educational institutions use web-based instruction because of the country's extensive use of information technology. With the aid of learning motivation and techniques, students' performance is predictable. Students who exhibit more adaptable self-regulatory mechanisms also show better learning efficacy and more motivation for learning (Liu & Yu, 2012). One of the most crucial 21st-century skills is digital competency. To use digital resources in the classroom more effectively, instructors need to be proficient in digital technologies. Given all of the advantages that digital competencies will bring to students both in the classroom and in the years to come, teacher education programs ought to incorporate them (Yesmakhanova et al., 2022). Higher education institutions are confronted with the problem of effectively incorporating new technology into their curricula in the twenty-first century. It is anticipated that e-learning will revolutionize the improvement of individualized learning (Almaiah et al., 2016b). Although e-learning has been around for a while, there is still discussion regarding educational values, such as the lack of personality education in the e-learning environment, which makes it seem like it is still in its infancy. Higher education institutions find that blended learning works better than only online or in-person instruction. A method of teaching that blends traditional in-person and virtual instruction is called blended learning. In a hybrid learning setting as opposed to an online or in-person learning environment, students learn more efficiently (Long & Van Hanh, 2020).

Online learning has grown in popularity in higher education as a result of the Internet and Education 4.0. However, we cannot discount the importance of conventional face-to-face learning, where the educational content may be efficiently conveyed through direct connection between professors and students, even in spite of certain fantastic features of online learning (such as flexibility in learning). Many colleges have begun using "blended learning," an innovative teaching approach, to capitalize on the benefits of multiple learning styles (Cuong et al., 2022). Online and blended learning environments are revolutionizing education by offering flexible, accessible, and personalized learning experiences. These modalities combine traditional face-to-face instruction with digital tools and resources, allowing students to engage with content at their own pace and convenience (Fernández-Batanero et al., 2022). Online learning provides opportunities for remote education, breaking geographical barriers and enabling lifelong learning (Edu et al., 2022). Blended learning integrates the best of both worlds, enhancing in-person instruction with online activities that promote deeper understanding and collaboration. Together, these innovative approaches are transforming the educational landscape, making learning more inclusive, interactive, and adaptive to the needs of diverse learners.

The educational system is evolving to integrate advancements in technology, aiming to foster critical thinking, problem-solving, and active student engagement. Blended learning, which combines online and in-person methods, is increasingly used to meet the needs of millennial learners who thrive in digital environments. As technology reshapes educational settings, web-based activities are now transforming classrooms into hybrid environments that blend online and face-to-face interactions. This shift is particularly evident in higher education, where blended learning has proven more effective than purely online or in-person approaches. In countries like China, web-based instruction is widely adopted, and digital competency has become a critical 21st-century skill. Teachers must now be proficient in digital tools to enhance classroom learning and better prepare students for the future. While online learning offers flexibility and removes geographical barriers, traditional face-to-face instruction remains vital for delivering educational content through direct interaction. Blended learning capitalizes on the strengths of both modalities, providing students with a more efficient and personalized learning experience. As education continues to evolve with innovations like Education 4.0, blended and online learning are making education more inclusive, accessible, and adaptive, transforming the learning landscape to cater to diverse student needs.

## OVERVIEW OF THE EVOLUTION OF ONLINE AND BLENDED LEARNING

The evolution of online and blended learning has been shaped by advancements in technology, educational theory, and the increasing need for flexible learning solutions. Initially, online learning was limited to simple text-based courses delivered via email or early Learning Management Systems (LMS). Over time, the integration of multimedia, interactive content, and communication tools like discussion forums and video conferencing enhanced the online learning experience. The rise of MOOCs (Massive Open Online Courses) in the early 2010s revolutionized access to education, offering courses from top universities to global learners. Blended learning, which combines traditional face-to-face instruction with online components, has also grown, providing the benefits of both models. During the COVID-19 epidemic, smart mobile learning (M-learning) applications demonstrated a number of novel advantages for higher education institutions. These applications were utilized to facilitate distance learning. (Almaiah et al., 2022). The COVID-19 pandemic accelerated this transition, prompting institutions worldwide to adopt digital learning tools on a larger scale. Today, online and blended learning are characterized by adaptive learning technologies, personalized learning pathways, and enhanced learner engagement through AI and data analytics, making education more accessible and tailored to individual needs. The evolution of blended learning approaches brought about by the changes in the distant education system has made adequate support necessary to help students study. When it comes to social support services, students' interactions with tutors and other students were found to be the most beneficial, while counseling services were not provided for these students (Amponsah et al., 2021). Blended learning encompasses not only the concept of using online technologies to complement classroom activities but also the merging of technology and traditional transmission means to facilitate learning, student, and instructor activities (Amponsah et al., 2021).

In the twenty-first century, online learning has grown in popularity, but during the COVID-19 worldwide pandemic in 2020, it became an extremely crucial mode of instruction (Xu et al., 2023). Because of the coronavirus disease (COVID-19) pandemic, online and blended learning have become increasingly significant parts of the teaching and learning processes at educational institutions around the world (Ali et al., 2023). In the modern world, it is not unusual for each of us to be connected to one or more communities at any given time. It permeates every aspect of our social interaction as fellow humans. Information retrieval and computer-to-computer and person-to-person communication are now easier than ever thanks to the Internet, an electronic network of computers. Many people think that the Internet has two sides because of its rapid expansion and the stories of knowledge dispersing across borders (Hanif et al., 2022).

Blended learning methods have been increasingly popular in recent years, eventually replacing traditional classroom instruction due to the rapid advancements in the Internet and educational technology. The hybrid learning environment stimulates students' initiative, enthusiasm, and creativity in learning by combining the benefits of networked online learning with traditional learning methods. Students can access learning resources and teachers' assistance equally anytime and anywhere (Jia, 2022). Education has been significantly influenced by the coronavirus disease (COVID-19) outbreak, among other aspects of life. Many educational institutions were forced to close as a result of the lockdown measures put in place after the COVID-19 virus spread over the planet (Ali et al., 2023). Amidst the COVID-19 pandemic, many universities are facing the primary problem of providing and utilizing online and e-learning systems. Blackboard and other e-learning platforms have a number of amazing capabilities that might be helpful in the current COVID-19 pandemic. During this pandemic, e-learning resources are essential. E-learning platforms can help educators plan, organize, carry out, and monitor the teaching and learning process (Almaiah et al., 2020). The transition to online and mixed learning has made it abundantly clear how crucial it is for students' motivation, emotions, and involvement in self-regulated learning (SRL) to be successful. The pandemic brought to light the necessity for students, especially those pursuing online education, to be able to control or regulate their emotions, such as loneliness or anxiety, and to maintain their motivation and focus in order to learn and succeed academically (Xu et al., 2023). In the context of blended learning, in-person settings are utilized to provide access for students, boost their motivation, and foster interpersonal relationships through in-class activities. Meanwhile, various internet platforms are employed in virtual settings for the same objectives. These online spaces include learning management systems (LMS), social networks, Web 2.0 applications, and MOOCs (Mese & Dursun, 2019). An unquestionable and useful instrument for everyone involved in the education spectrum, online education has solidly established a basis in education. The emphasis of the educational community has shifted to the blended classroom, which combines virtual learning with traditional, in-person instruction, to give special needs kids the best possible chances. The efficacy of blended learning environments in teaching fundamental concepts and enhancing the comprehensive evaluation of children with special needs and learning difficulties has been established (Rivera, 2017).

Using both in-person and virtual learning settings together creates blended learning environments. Benefiting from both in-person and virtual learning settings is possible for educators and students when blended learning environments are utilized (Yesmakhanova et al., 2022). Blended learning gives students flexibility and choice in their education. The students are free to select when and where they want to engage in the online course. It also demonstrates that self-efficacy in blended learning is higher than in other teaching modalities. Additionally, students who take blended learning courses perform better than those who take fully online or in-person classes (Puspitasari, 2021). The disruptive era, which jeopardizes technological advancement, proposes a new paradigm for the teaching-learning procedure. The term for it is blended learning. So, as part of the current teaching trends, educators implement it in their classrooms (Puspitasari, 2021). Students are encouraged to consider the information from a variety of angles by using this approach. Students can develop their online learning skills in addition to sitting in class (Puspitasari, 2021).

The evolution of online and blended learning has been driven by technological advancements, educational theory, and the growing demand for flexible learning solutions. Initially, online learning began with simple text-based courses but has evolved to incorporate multimedia, interactive content, and communication tools that enhance the learning experience. The rise of MOOCs expanded global access to education, while blended learning merged online and traditional methods, offering the best of both worlds. The COVID-19 pandemic accelerated the adoption of digital learning tools, making online and blended learning essential to the educational process. These approaches now feature personalized learning paths, AI-driven adaptive technologies, and increased student engagement. Blended learning combines in-person and virtual environments, allowing students access to learning resources anytime, anywhere, while fostering motivation and interpersonal connections. This model has become particularly valuable in addressing diverse student needs, including those with special learning requirements. Blended learning also encourages self-regulation, creativity, and critical thinking, enabling students to manage their learning at their own pace and in various settings. As the educational landscape continues to shift, blended learning is emerging as a crucial pedagogical approach, offering flexibility, improved student performance, and a balanced combination of online and face-to-face interactions.

## THEORETICAL FRAMEWORK: THEORIES UNDERPINNING ONLINE AND BLENDED LEARNING

The theoretical framework underpinning online and blended learning draws from several key educational theories that inform the design and delivery of these learning models. Constructivist theory, notably from Piaget and Vygotsky, emphasizes the learner's active role in constructing knowledge through interaction with content, peers, and instructors. Online learning environments foster this through collaborative activities, discussion forums, and problem-based learning, encouraging learners to engage deeply with materials and build understanding in a social context. Connectivism, introduced by Siemens and Downes, is particularly relevant to online and blended learning, positing that knowledge resides in networks, and learning is the ability to navigate and synthesize information from diverse sources. This theory aligns with the nature of digital learning environments, where learners access vast online resources and engage in knowledge-building communities. Cognitive load theory, as proposed by Sweller, also plays a role, especially in the design of online content, which must be structured to optimize learners' cognitive processing. Additionally, Garrison, Anderson, and Archer's Community of Inquiry (CoI) framework is pivotal in online and blended learning, focusing on the development of social, cognitive, and teaching presence to create an engaging and effective learning experience. Behaviorism also influences online learning, particularly in the use of quizzes, gamification, and immediate feedback mechanisms, reinforcing learning behaviors. Overall, these theories guide the creation of learner-centered, interactive, and adaptive learning environments in both online and blended formats.Blended learning is characterized by the amalgamation of online and classroom learning, combining the benefits and attributes of both settings. It is important to consider the characteristics of the users while developing a mixed-learning environment. Stated differently, users' learning requirements in this environment vary depending on their particular qualities, learning preferences, and past knowledge (Yesmakhanova et al.,

2022). The planning, creation, and application of technology-mediated learning environments depend critically on the motivation of students to learn. In a constructivist-based blended learning environment (CBLE), engagement learning strategies have been developed to support students' learning (Machumu et al., 2018). In actuality, blended learning satisfies learners' needs for access to learning resources at any time and from any location. With this approach, a significant quantity of theoretical information is disseminated via the Internet, significantly enhancing the in-person interactions between educators and learners (Cuong et al., 2022).

Powerful elements of both in-person and virtual learning have given rise to the idea of blended learning. Blended learning involves students acquiring knowledge primarily through online means and supplementing it with in-person instruction at a supervised site away from their residence (Mese & Dursun, 2019). With its open platform for discussion between instructor and student, the conventional classroom offers a traditional method of instruction. The use of technology and online additional resources, intended to engage students at all academic levels, has created new educational opportunities within the past 25 years (Rivera, 2017). The idea of a blended classroom has shown to be effective in accomplishing the objective of providing special education students with learning problems with a constructive, meaningful learning experience. This goal is shared by general education and special needs teachers. Special needs kids can benefit from both virtual resources and having a live, in-person teacher to ask questions and get advice from in a blended learning environment (Rivera, 2017).

Through careful planning and the use of relevant tools, educators can efficiently offer learning materials through the use of digital and web-enhanced teaching platforms, which are used by many educational administrations worldwide. The novel hybrid approach known as blended learning combines traditional in-person or FTF learning with online and physical learning, as well as concurrent delivery of instructional materials via the internet (Ali et al., 2023). Blended learning is a comprehensive approach that includes online instruction and learning. It efficiently combines various modes of distributing learning materials, teaching models, and learning styles to create a methodical and intentional approach that improves first-time factor engagement rates by using relevant technology

(Ali et al., 2023). Different learning strategies should be introduced and described in English teaching and learning to cater to the diverse learning motivations of students. This would help to close the efficacy gap in students' English learning in a blended learning environment (Liu & Yu, 2012).

The theoretical framework underpinning online and blended learning is informed by several educational theories that shape how these learning environments are designed and implemented. Constructivist theory, particularly from Piaget and Vygotsky, emphasizes the active role of learners in constructing their knowledge through interactions with content, peers, and instructors. In online and blended learning, this theory is realized through collaborative activities, discussions, and problem-based learning that encourage students to engage deeply and build understanding within a social context. Connectivism, introduced by Siemens and Downes, further supports these models by highlighting the importance of navigating and synthesizing information from various digital sources, as learning is viewed as the ability to tap into networks of knowledge. This is especially relevant in online environments where vast resources and knowledge communities are accessible. Cognitive load theory by Sweller plays a crucial role in the design of online learning content, ensuring that materials are structured to optimize learners' cognitive processes and avoid overwhelming them. Additionally, the Community of Inquiry (CoI) framework by Garrison, Anderson, and Archer emphasizes the importance of social, cognitive, and teaching presence in creating a meaningful and engaging learning experience. It highlights the need for interaction, critical thinking, and effective instruction in online settings. Behaviorism, too, influences online learning through the use of quizzes, gamification, and instant feedback, reinforcing learning behaviors and engagement. Together, these theories guide the development of learner-centered, adaptive, and interactive environments in both online and blended learning contexts, offering a balanced approach to modern education that caters to the needs of diverse learners.

## TECHNOLOGICAL INNOVATIONS: IMPACT OF TECHNOLOGY ON TEACHING AND LEARNING

Technological innovations have profoundly impacted teaching and learning, transforming the educational landscape by enhancing accessibility, engagement, and personalization. The integration of technology in classrooms, from basic tools like computers and projectors to advanced digital platforms and mobile applications, has made learning more interactive and dynamic. Learning Management Systems (LMS) such as Moodle and Blackboard allow educators to organize and deliver content, assess students, and facilitate collaboration remotely. The rise of artificial intelligence (AI) and machine learning has enabled the development of adaptive learning systems, which tailor content to individual learners' pace, preferences, and performance, offering a personalized learning experience. Virtual Reality (VR) and Augmented Reality (AR) technologies are bringing immersive, hands-on experiences to subjects like science, history, and engineering, allowing students to explore complex concepts in a simulated environment. The use of video conferencing tools like Zoom and Microsoft Teams has facilitated real-time, synchronous learning, making education accessible globally, especially during the COVID-19 pandemic. Moreover, the use of data analytics helps educators track student progress, identify learning gaps, and intervene early to provide support. Digital collaboration tools, such as Google Classroom and Microsoft OneNote, enhance peer interaction and group learning, breaking down geographical barriers. Overall, technology has democratized education by providing flexible learning paths, fostering collaborative learning environments, and ensuring continuous access to educational resources, revolutionizing the traditional model of teaching and learning. Information and communication technology (ICT) integration into the educational setting has increased in frequency in recent years. The idea of blended learning (BL), which combines in-person instruction with online contact between teachers and students, is introduced to most disciplines alongside e-learning, or learning through technology. Although the general satisfaction with theory and real practices is still low, the goal is to prepare students for self-directed learning (Seman et al., 2019). Modern information and communication technologies that mimic some characteristics of human thought processes will play a crucial role in the development of blended learning. This implies that educators must continuously enhance their digital competencies in order to function well in a blended learning setting (Chervonyi et al., 2021). The primary reason for the acknowledged requirement for professional development of ICT skills is the acquisition of new technologies required in a blended learning environment. For blended learning teachers, for instance, the ideal technology answer is to employ a learning management system (Chervonyi et al., 2021).

Technological innovations have significantly transformed teaching and learning, enhancing accessibility, engagement, and personalization within education. The integration of tools ranging from basic computers to advanced digital platforms has made learning more interactive and dynamic. Learning Management Systems (LMS) like Moodle and Blackboard streamline content delivery, assessment, and collaboration, enabling remote education. The emergence of artificial intelligence (AI) and machine learning has led to adaptive learning systems that customize educational experiences based on individual learners' needs and performance, fostering personalized learning. Immersive technologies like Virtual Reality (VR) and Augmented Reality (AR) provide hands-on experiences in subjects such as science and history, allowing students to explore complex concepts in simulated environments. The use of video conferencing tools, particularly highlighted during the COVID-19 pandemic, has facilitated real-time learning, making education globally accessible. Data analytics empower educators to monitor student progress and identify learning gaps, enabling timely interventions. Digital collaboration tools, such as Google Classroom and Microsoft OneNote, promote peer interaction and group learning while breaking geographical barriers. Overall, technology has democratized education by offering flexible learning paths and fostering collaborative environments, revolutionizing traditional teaching methods. As the integration of Information and Communication Technology (ICT) continues to grow, blended learning, a combination of in-person and online interactions, has become increasingly prominent. However, for this approach to be effective, educators must continuously enhance their digital competencies, ensuring they are equipped to navigate and leverage new technologies effectively within the blended learning landscape.

## PEDAGOGICAL STRATEGIES: EFFECTIVE ONLINE AND BLENDED TEACHING METHODOLOGIES

Effective online and blended teaching methodologies require pedagogical strategies that leverage the strengths of both digital and face-to-face learning environments to enhance student engagement, interaction, and mastery of content. One key approach is the flipped classroom model, where students engage with lecture materials, readings, or multimedia content online before class and use in-person or synchronous sessions for active learning through discussions, problem-solving, and collaborative projects. Active learning strategies, such as case studies, simulations, and group work, are essential in both online and blended formats, encouraging students to apply knowledge in real-world scenarios while interacting with peers and instructors. The use of scaffolded learning is another crucial methodology, where complex concepts are broken down into smaller, manageable tasks, allowing students to build knowledge progressively with constant feedback. Asynchronous learning, facilitated through discussion boards, recorded lectures, and self-paced assignments, empowers learners to take control of their learning process, while synchronous sessions create opportunities for real-time interaction and immediate clarification of doubts. Instructors also use multimodal content delivery, combining text, video, audio, and interactive media, to cater to diverse learning preferences, making material more accessible and engaging. Formative assessments are frequently used to provide timely feedback and keep learners on track, while peer assessments foster collaboration and deeper learning. Additionally, the incorporation of gamification elements such as quizzes, badges, and leaderboards motivates learners by adding an element of fun and competition. A learner-centered approach, grounded in constructivist and connectivist theories, underpins these methodologies, emphasizing collaboration, critical thinking, and the practical application of knowledge. For the purpose of motivating pupils to engage in blended learning activities, teachers need to possess both confidence and creativity. Therefore, in order to generate great learning experiences, blended learning sessions should be loaded with fun and interactive activities rather than merely having people communicate through forums or post exercises to be done. Rather of depending solely on text or PowerPoint slides, the course should also include a variety of digital formats, such as animation and video. To develop a more innovative teaching approach, it's also a good idea to integrate different online teaching resources. Meanwhile, administrators in higher education should make sure the system is packed with features and resources to enable teachers and students to be as creative as possible (Seman et al., 2019).

Diversity in learner characteristics (such as individual differences) and diversity in online instructional delivery modes become important areas for research to ensure educational excellence, educational equity, and excellence for all learners as online and blended learning become more and more integrated into formal, K–12 as well as higher education systems. For instance, our scoping review shows that interventions, scaffolds, and integrated SRL technologies provide a variety of approaches to meet the learning needs of students (Xu et al., 2023). Since intrinsic motives have a beneficial effect on their drive to do so in a blended learning environment, teachers or lecturers could design an English learning environment that could inspire aspiring EFL teachers to practise writing essays (Apoko, 2023). With regard to their musical abilities, students' attitudes towards autonomous learning have increased with the introduction of information and communication technology (Ruokonen & Ruismäki, 2016).

Effective online and blended teaching methodologies harness the strengths of both digital and face-to-face learning to boost student engagement, interaction, and mastery of content. One prominent approach is the flipped classroom model, where students review lecture materials or multimedia content online before attending class, allowing in-person sessions to focus on active learning through discussions and collaborative projects. Active learning strategies—such as case studies, simulations, and group work encourage students to apply knowledge in real-world contexts while interacting with peers and instructors. Scaffolded learning breaks complex concepts into smaller, manageable tasks, enabling gradual knowledge building with continuous feedback. Asynchronous learning, facilitated by discussion boards and self-paced assignments, empowers learners to control their learning process, while synchronous sessions provide real-time interaction and immediate clarification. Instructors use multimodal content delivery, combining text, video, audio, and interactive media to cater to diverse learning preferences and enhance accessibility.

Formative assessments offer timely feedback, while peer assessments foster collaboration and deeper understanding. Incorporating gamification elements like quizzes and leaderboards adds motivation through fun and competition. These methodologies are grounded in learner-centered approaches, emphasizing collaboration, critical thinking, and practical application of knowledge. For successful blended learning experiences, educators must exhibit confidence and creativity, ensuring sessions are rich with interactive activities rather than relying solely on forums or static presentations. Integrating varied digital formats such as animations and videos—can enhance engagement, while administrative support in higher education is crucial for fostering creativity among educators and students. As online and blended learning become increasingly integrated into K-12 and higher education, addressing diversity in learner characteristics and instructional delivery modes is vital for ensuring educational equity and excellence. Research shows that interventions, scaffolds, and self-regulated learning (SRL) technologies can effectively meet diverse learning needs. Additionally, fostering intrinsic motivation in students, particularly in areas like English language learning, can inspire them to engage more deeply in their educational journeys. Overall, embracing innovative teaching strategies and leveraging technology can significantly enhance the learning experience for all students.

## BENEFITS AND CHALLENGES: ADVANTAGES OF ONLINE AND BLENDED LEARNING (FLEXIBILITY, ACCESSIBILITY, PERSONALIZED LEARNING, ETC.)

Online and blended learning offers a range of advantages that make education more flexible, accessible, and personalized, meeting the diverse needs of today's learners. One of the most significant benefits is flexibility, as students can access course materials, lectures, and assignments at their own pace and schedule, allowing them to balance education with work, family, and other responsibilities. This is particularly valuable for adult learners, working professionals, and those in remote areas where traditional in-person education may be challenging. Accessibility is another key advantage, as online learning eliminates geographical barriers, enabling learners from different parts of the world to enroll in courses or degree programs that may not be available locally. It also makes education more inclusive for students with disabilities, offering adaptive technologies, such as screen readers, subtitles, and adjustable content formats, to accommodate various learning needs. The use of personalized learning platforms, powered by AI and data analytics, tailors the educational experience to each student's strengths, weaknesses, and learning preferences, providing customized content, pacing, and feedback. This individualized approach enhances engagement and allows students to focus on areas where they need the most improvement. Cost-effectiveness is another benefit, as online courses often reduce or eliminate the need for commuting, housing, and physical materials, making education more affordable. Furthermore, collaborative tools like discussion forums, virtual breakout rooms, and group projects foster peer interaction, enhancing the social dimension of learning despite physical distance. However, these benefits also come with challenges, such as the need for self-discipline and time management, digital literacy, and consistent access to reliable technology and the Internet. Despite these hurdles, online and blended learning offers a dynamic, learner-centered approach that is increasingly recognized as an essential part of modern education. Online learning platforms are useful for improving the academic and practical learning experiences of students. Additionally, you can leverage online elements to improve student learning, such as threaded discussions, assessment tools for tasks, and centralized course management systems for course resources. In mixed-learning environments, implementing blended learning may be an effective pedagogical strategy for teaching and learning (Yuan, 2023). In today's heavily tech-influenced school, blended learning a mix of traditional classroom instruction and online learning is the perfect pedagogical strategy for student learning. While using a blended learning strategy can improve students' learning outcomes, there are drawbacks to using technology and the Internet, such as technical difficulties with online learning platforms. To successfully implement blended learning as an educational strategy for efficient instruction and learning (Yuan, 2023). Higher education institutions' (HLIs') continuous study on student motivational variables and engagement tactics in authentic learning activities is still very important. A high degree of student involvement in purposeful, long-term learning experiences is necessary for the adoption of innovative instructional technology, curriculum redesign, and constructivist-based blended learning environments (CBLE) (Machumu et al., 2018).

There are certain advantages to blended learning for both educators and students. Blended learning offers learners the flexibility to work at their own pace without interfering with other learners. Students can learn whenever they want, wherever they are, and at their own pace because of its adaptability. Additionally, it helps teachers adjust to students' varied learning styles more skillfully (Puspitasari, 2021). Additionally, blended learning allows students to focus more on themselves. It makes it possible for teachers to provide every student with the attention they need. It is achieved by creating a course that pays closer attention to the demands of the learners. It is urged learners to go beyond the requirements of the course and broaden their learning. They will go over and reaffirm everything that was covered in class online, at their own pace (Puspitasari, 2021). Because blended learning settings are convenient and flexible, (Roff, 2018) said, the students were happy with them. Technology and separation were two areas where people expressed displeasure. Self-determined practices are an excellent way for parents and instructors to support their online learners. To encourage students' autonomous (also known as identified or intrinsic) motivation, instructors should design learning activities that give them more options and prompt them to make thoughtful or deliberate decisions about what, when, and how they would learn in their classes (Xu et al., 2023). It may be feasible to use and execute mobile learning services, but there may be challenges to this technology's success because it might not satisfy students' needs and expectations (Almaiah et al., 2016a).

Working in teams during blended learning, which includes online collaborations and various group work techniques, is one of the most efficient ways to enhance social competency in FLT. However, in FLT, group work is a prerequisite for effective communication and interaction, and consequently for the effective development of social competence (Prohorets & Plekhanova, 2015). An administrative standpoint views blended learning as a chance to lower running expenses, expand access to the institution's educational programs, and improve the higher education institution's reputation. The difficulties with this learning form are that it must be in line with institutional priorities and goals; additionally, organizational change is met with opposition; and lastly, there is a dearth of organizational structure, collaboration, and partnership experiences (Chervonyi et al., 2021). With blended learning, there are more options for educators and students to advance their careers. This occurs when students learn outside of their homes in a supervised brick-and-mortar setting. It means that the students go to a real school with instructors or mentors. Put another way, while students are learning in a variety of ways, the online and in-person components combine to offer an integrated course (Puspitasari, 2021). One of the major technological advancements in modern education that has the potential to address numerous learning problems is blended learning environments. Blended learning is still in its infancy, hence further study is needed to identify more effective models and designs for blended learning initiatives in educational settings (Al-ismaiel, 2022). Due to the aforementioned drawbacks, blended learning generally appears to be flawed despite offering a suitable setting for an interactive and customised teaching approach. Therefore, in order to make the courses exciting and engaging, interactive solutions are required. Additionally, experiential learning methodologies must be implemented in accordance with the technical learning environment's characteristics (Cuong et al., 2022).

Online and blended learning provides numerous advantages, making education more flexible, accessible, and personalized for today's diverse learners. One of the primary benefits is flexibility, allowing students to access course materials at their own pace, which is particularly beneficial for adult learners and those in remote areas. This mode of education eliminates geographical barriers, making it inclusive for students with disabilities and providing adaptive technologies to accommodate various learning needs. Personalized learning platforms leverage AI and data analytics to tailor the educational experience, enhancing engagement by focusing on individual strengths and weaknesses. Additionally, blended learning reduces costs by minimizing commuting and material expenses. It fosters peer interaction through collaborative tools, enhancing the social aspect of learning despite physical distance. However, challenges such as the need for self-discipline, digital literacy, and consistent access to technology must be addressed. While blended learning can significantly improve student outcomes, technical difficulties can hinder its effectiveness. For educators, this approach allows for greater adaptability to diverse learning styles and facilitates personalized attention to students, encouraging them to engage more deeply with the material. Despite some dissatisfaction with technology, the overall convenience and flexibility of blended learning environments lead to higher student satisfaction. Supporting autonomous learning through thoughtful activity design can further enhance motivation and engagement. Additionally, group work in blended settings promotes social competence and communication skills. Administratively, blended learning can reduce operational costs and enhance institutional reputation, although it requires alignment with organizational goals and may face resistance to change. Ultimately, blended learning presents opportunities for students and educators to expand their capabilities through integrated learning experiences, combining online and in-person components. As this approach continues to evolve, ongoing research is essential to develop more effective models and designs, ensuring that blended learning remains engaging and interactive while addressing potential drawbacks.

## ASSESSMENT AND EVALUATION: APPROACHES TO ASSESSING STUDENT LEARNING IN ONLINE AND BLENDED ENVIRONMENTS

Assessing student learning in online and blended environments requires a combination of formative and summative assessment approaches that are adaptable to the unique characteristics of digital learning. Formative assessments play a crucial role, offering ongoing feedback through quizzes, discussion posts, peer assessments, reflective journals, and low-stakes assignments that help students track their progress and identify areas for improvement. These assessments also allow instructors to monitor student engagement and comprehension, enabling timely interventions. Summative assessments, such as final exams, projects, or research papers, are often administered at the end of a course or module and can be conducted through online platforms using tools that ensure academic integrity, like proctoring software or plagiarism detection systems. One emerging trend is the use of authentic assessments, which require students to apply their knowledge in real-world contexts through problem-solving tasks, case studies, simulations, or digital portfolios that showcase their skills over time. In blended environments, assessments may combine in-person presentations with online submissions, leveraging both environments to evaluate students comprehensively. Peer assessment is another effective strategy, promoting collaboration and critical thinking by allowing students to evaluate each other's work based on established rubrics. Additionally, self-assessment encourages metacognitive reflection, where students assess their learning journey and identify personal growth areas. The use of data analytics is also transforming assessment, with learning management systems (LMS) tracking student performance and offering insights into areas of difficulty, allowing for personalized feedback. Rubrics and clear guidelines are essential in ensuring transparency and fairness in both online and blended assessments, while e-portfolios provide a holistic view of student development by capturing diverse artifacts of learning over time. Together, these approaches create a comprehensive assessment framework that supports continuous learning and ensures that both process and product are evaluated effectively. Blended learning approaches have become more prevalent as a result of the changes made to the distance education system, and this has made effective assistance necessary to help students learn. Regarding social support services, it was found that students could primarily rely on their interactions with tutors and fellow students; counseling services were not provided to these students (Amponsah et al., 2021). More possibilities for skill improvement are given to educators and students through blended learning. It takes place while students study away from home in a supervised brick-and-mortar setting. It suggests that the students attend a real school with instructors or mentors. Stated differently, learners engage in a mixed-methods learning environment as the online and in-person components complement each other to create an integrated course (Puspitasari, 2021). High-quality, methodical research is required to guide instructional design and methodology to proceed successfully and take advantage of all that blended and online learning has to offer students (Xu et al., 2023).

Assessing student learning in online and blended environments necessitates a blend of formative and summative assessment strategies tailored to the distinct nature of digital learning. Formative assessments are vital, providing continuous feedback through quizzes, discussion posts, peer assessments, and reflective journals, which help students monitor their progress and pinpoint areas for improvement. These assessments also enable instructors to gauge student engagement and comprehension, facilitating timely interventions. Summative assessments, conducted at the end of a course or module, can be administered online using tools like proctoring software to ensure academic integrity. An emerging trend is authentic assessments, which require students to apply their knowledge in real-world scenarios through problem-solving tasks, case studies, and digital portfolios. In blended environments, assessments often combine in-person presentations with online submissions, providing a comprehensive evaluation of student performance. Peer assessments promote collaboration and critical thinking by allowing students to evaluate each other's work, while self-assessment fosters metacognitive reflection. Data analytics from learning management systems enhance the assessment process by tracking student performance and offering insights for personalized feedback. Clear rubrics and guidelines ensure transparency and fairness, while e-portfolios capture a holistic view of student development over time. Together, these methods create a robust assessment framework that supports continuous learning and evaluates both processes and outcomes effectively. As blended learning becomes more common, it highlights the need for quality research to inform instructional design and methodologies, ultimately enhancing the educational experience for students and educators alike.

## FUTURE DIRECTIONS: EMERGING TRENDS AND FUTURE POSSIBILITIES

Emerging trends and future possibilities in online and blended learning are being shaped by rapid advancements in technology, evolving educational needs, and a growing emphasis on personalized and flexible learning. One major trend is the increasing use of artificial intelligence (AI) to create adaptive learning systems that can personalize educational experiences based on students' performance and learning styles (Shishakly et al., 2023). These systems analyze data to predict learning paths, offering real-time feedback and customized resources. Augmented Reality (AR) and Virtual Reality (VR) are also set to play a bigger role, providing immersive and experiential learning opportunities, especially in fields like medicine, engineering, and history, where students can interact with complex simulations in safe, virtual environments. The rise of microlearning, where content is delivered in small, easily digestible modules, reflects the growing need for just-in-time learning, especially for professionals seeking to update skills. Blockchain technology is another emerging trend, promising to enhance the credibility and security of digital certifications and student records. As education becomes more global, the trend toward global classrooms will expand, with students from different countries collaborating through virtual platforms, creating culturally diverse learning experiences. Additionally, the growing focus on equity and inclusion will drive innovations that make learning more accessible for students with disabilities, using assistive technologies and adaptive content delivery methods. Lifelong learning will continue to gain importance, with institutions offering more flexible, modular programs that cater to the needs of adult learners and professionals. As educational technology continues to evolve, the lines between traditional, online, and blended learning will blur, creating a more fluid and hybrid learning environment that is student-centered, dynamic, and responsive to the changing demands of the global workforce. Providing an effective method will surely lead to a higher satisfaction percentage. As a result of their great and welcoming blended learning experience, students will stick with the platform. Meanwhile, the study's results cannot be broadly applied, despite the fact that the empirical findings have added to the body of literature already in existence for a distinct populace. Therefore, research in the future ought to be done in various educational settings. Furthermore, it is imperative that future research studies assess the extent to which the blended learning system enhances students' performance within a mixed-learning setting (Seman et al., 2019). Because instructors and students can successfully utilize e-learning if institutions are constantly updating their technical resources, the administration of the university must supply the required hardware, software, and internet connection (Almaiah et al., 2020).

There is an urgent need to reconsider topics like interaction types and intensities in the electronic environment versus traditional classroom interactions when thinking about blended learning in some foreign language teaching (FLT) scenarios (Prohorets & Plekhanova, 2015). The development of social competence in a blended learning setting differs slightly from a typical classroom, which is something that instructors and course designers in blended learning should have in mind. The development of social competence is aided by the technological, didactic, and methodological resources of blended learning; yet, not all educators can fully realize this potential (Prohorets & Plekhanova, 2015). One of the key competencies in the current world is social competence, and group work, as a pedagogical method used in FLT, has a positive impact on the development of social competence. Blended learning has the potential to enhance social skills as a result of its technological, didactic, and methodological resources (Prohorets & Plekhanova, 2015). The intermediate level encompasses both human and non-human interaction qualities since it integrates social and technological interaction components. While they interact asynchronously, students receive feedback asynchronously based on how other participants manipulate the course's virtual content interactively. Students' engagement in peer-to-peer collaborative activities in both in-person and virtual settings is prioritized (Prohorets & Plekhanova, 2015). It is important to look into different teaching methods in online learning environments to help students in public education improve their critical thinking abilities. Reviewing the perspectives of educators and members of the learning staff regarding the advantages and disadvantages of using blended learning environments in the classroom is also recommended (Alismaiel, 2022).

Emerging trends and future possibilities in online and blended learning are being significantly influenced by rapid technological advancements, shifting educational needs, and a growing focus on personalized, flexible learning experiences. A prominent trend is the increased integration of artificial intelligence (AI) to develop adaptive learning systems that tailor educational experiences to individual students' performance and learning styles. These systems analyze data to forecast learning paths, providing real-time feedback and customized resources. Augmented Reality (AR) and Virtual Reality (VR) are expected to enhance immersive learning, particularly in fields like medicine and engineering, where students can safely interact with complex simulations. The rise of microlearning, characterized by small, easily digestible content modules, reflects the demand for just-in-time learning, especially among professionals looking to upgrade their skills. Blockchain technology also emerges as a trend, promising improved security and credibility for digital certifications and student records.

Global classrooms are expanding as education becomes more international, allowing students from diverse backgrounds to collaborate virtually. The emphasis on equity and inclusion is driving innovations that enhance accessibility for students with disabilities, utilizing assistive technologies and adaptive content delivery. Lifelong learning will gain prominence, with institutions offering flexible, modular programs catering to adult learners and professionals. As educational technology continues to evolve, the distinctions between traditional, online, and blended learning will blur, fostering a more dynamic, student-centered, and responsive learning environment. In this context, there is a need for research to assess the effectiveness of blended learning systems in various educational settings. Future studies should explore the impact of interaction types and intensities in electronic versus traditional classroom settings, particularly in foreign language teaching. The development of social competence is also essential, as blended learning can enhance social skills through its technological and pedagogical resources, although not all educators are fully equipped to leverage this potential. Investigating diverse teaching methods in online environments is crucial for improving students' critical thinking abilities, while gathering insights from educators regarding the advantages and disadvantages of blended learning will further inform effective practices in the classroom.

## CONCLUSION

In conclusion, the evolving landscape of online and blended learning is being shaped by technological advancements and an increasing emphasis on personalized education, flexibility, and accessibility. The integration of artificial intelligence, AR, and VR is enhancing immersive and adaptive learning experiences, while trends such as microlearning and blockchain technology promise to streamline content delivery and secure student credentials. The shift towards global classrooms fosters diverse learning experiences, promoting collaboration among students from different backgrounds. Moreover, the focus on equity and inclusion is vital for ensuring that educational resources are accessible to all learners, including those with disabilities. As lifelong learning becomes more prominent, educational institutions are adapting to meet the needs of adult learners through flexible and modular programs. However, to fully realize the potential of blended learning, ongoing research is necessary to evaluate its effectiveness across various contexts and identify best practices. This includes examining interaction dynamics in both electronic and traditional environments and understanding how blended learning can facilitate the development of critical thinking and social competence among students. By addressing these areas, educators can enhance the learning experience, fostering a more dynamic and responsive educational landscape that meets the diverse needs of today's learners. Professional development for teachers in online and blended learning environments is essential to ensure effective instructional practices and maximize student engagement. As educational technology evolves, educators must enhance their digital competencies, familiarizing themselves with various learning management systems, online collaboration tools, and assessment strategies tailored for digital contexts. Continuous training should focus on integrating pedagogical theories with technology, enabling teachers to create learner-centered experiences that accommodate diverse learning styles and preferences. Workshops, online courses, and collaborative learning communities can provide educators with the necessary skills to design interactive content, utilize data analytics for personalized learning, and implement effective assessment practices. Additionally, fostering a growth mindset among educators encourages innovation and adaptability in their teaching approaches. By investing in comprehensive professional development, educational institutions can empower teachers to thrive in online and blended learning settings, ultimately enhancing the overall quality of education.

## REFERENCES

Akyüz, H. I., & Samsa, S. (2009). The effects of a blended learning environment on the critical thinking skills of students. *Procedia: Social and Behavioral Sciences*, *1*(1), 1744–1748. Advance online publication. DOI: 10.1016/j.sbspro.2009.01.308

Ali, A., Khan, R. M. I., & Alouraini, A. (2023). A Comparative Study on the Impact of Online and Blended Learning. *SAGE Open*, *13*(1), 1–10. DOI: 10.1177/21582440231154417

Alismaiel, O. A. (2022). Develop a New Model to Measure the Blended Learning Environments Through Students' Cognitive Presence and Critical Thinking Skills. *International Journal of Emerging Technologies in Learning*, *17*(12), 150–169. Advance online publication. DOI: 10.3991/ijet.v17i12.30141

Almaiah, M. A., Al-Khasawneh, A., & Althunibat, A. (2020). Exploring the critical challenges and factors influencing the E-learning system usage during COVID-19 pandemic. *Education and Information Technologies*, *25*(6), 5261–5280. DOI: 10.1007/s10639-020-10219-y PMID: 32837229

Almaiah, M. A., Ayouni, S., Hajjej, F., Lutfi, A., Almomani, O., & Awad, A. B. (2022). Smart Mobile Learning Success Model for Higher Educational Institutions in the Context of the COVID-19 Pandemic. *Electronics (Basel)*, *11*(8), 1–13. DOI: 10.3390/electronics11081278

Almaiah, M. A., Jalil, M. A., & Man, M. (2016a). Extending the TAM to examine the effects of quality features on mobile learning acceptance. *Journal of Computers in Education*, *3*(4), 453–485. DOI: 10.1007/s40692-016-0074-1

Almaiah, M. A., Jalil, M. A., & Man, M. (2016b). Preliminary study for exploring the major problems and activities of mobile learning system: A case study of Jordan. *Journal of Theoretical and Applied Information Technology*, *93*(2), 580–594.

Amponsah, S., Ussher, Y., & Benjamin, K. A. (2021). Availability and access to support services in a blended learning environment. *International Journal of Information and Communication Technology Education, 17*(1), 57–71. Advance online publication. DOI: 10.4018/IJICTE.2021010104

Apoko, T. W. (2023). EFL Preservice Teachers' Intrinsic Motivation in Writing Essays in a Blended Learning Environment. *VELES: Voices of English Language Education Society, 7*(1), 126–134. DOI: 10.29408/veles.v7i1.7838

Chervonyi, P., Alieksieieva, S., Senchylo-Tatlilioglu, N., Chastnyk, O., & Diomidova, N. (2021). Peculiarities of the higher education development in a blended learning environment. *Revista on Line de Política e Gestão Educacional, 25*(3), 2241–2257. DOI: 10.22633/rpge.v25i3.15959

Cuong, L. H., Thi, N., Giang, H., Thi, N., & Tu, T. (2022). Design of an Educational Game for Blended Learning Environment. *JST: Smart Systems and Devices, 32*(3), 17–25. DOI: 10.51316/jst.160.ssad.2022.32.3.3

Edu, T., Negricea, C., Zaharia, R., & Zaharia, R. M. (2022). Factors influencing student transition to online education in the COVID 19 pandemic lockdown: Evidence from Romania. *Ekonomska Istrazivanja, 35*(1), 3291–3304. DOI: 10.1080/1331677X.2021.1990782

Fernández-Batanero, J. M., Montenegro-Rueda, M., Fernández-Cerero, J., & Tadeu, P. (2022). Online education in higher education: emerging solutions in crisis times. In *Heliyon* (Vol. 8, Issue 8). DOI: 10.1016/j.heliyon.2022.e10139

Hanif, H., Hamzah, M., & Harun, R. N. S. R. (2022). Optimizing Participation with Helpers to Enhance Interaction in a Blended Learning Environment. *International Journal of Emerging Technologies in Learning, 17*(17), 155–170. Advance online publication. DOI: 10.3991/ijet.v17i17.26415

Jia, L. (2022). Clustering Analysis of the Evaluation Data on Students' Learning under a Blended Environment. *International Journal of Emerging Technologies in Learning, 17*(23), 128–144. Advance online publication. DOI: 10.3991/ijet.v17i23.35941

Liu, W., & Yu, H. (2012). Effectiveness study of English learning in blended learning environment. *Theory and Practice in Language Studies*, *2*(3), 524–530. DOI: 10.4304/tpls.2.3.524-530

Long, N. T., & Van Hanh, N. (2020). A structural equation model of blended learning culture in the classroom. *International Journal of Higher Education*, *9*(4), 99. Advance online publication. DOI: 10.5430/ijhe.v9n4p99

Machumu, H., Zhu, C., & Almasi, M. (2018). Students' motivational factors and engagement strategies in constructivist- based blended learning environments. *Afrika Focus*, *31*(1), 13–34. DOI: 10.1163/2031356X-03101002

Mese, C., & Dursun, O. O. (2019). Effectiveness of gamification elements in blended learning environments. *Turkish Online Journal of Distance Education*, *20*(3), 119–142. DOI: 10.17718/tojde.601914

Prohorets, E., & Plekhanova, M. (2015). Interaction intensity levels in blended learning environment. *Procedia: Social and Behavioral Sciences*, *174*(3822), 3818–3823. DOI: 10.1016/j.sbspro.2015.01.1119

Puspitasari, D. (2021). BLENDED LEARNING ENVIRONMENT: PROMOTING LEARNERS' SELF-EFFICACY. *ELTALL: English Language Teaching. Applied Linguistic and Literature*, *2*(1), 39–46. DOI: 10.21154/eltall.v2i1.2729

Rivera, J. H. (2017). The Blended Learning Environment: A Viable Alternative for Special Needs Students. *Journal of Education and Training Studies*, *5*(2), 79–84. DOI: 10.11114/jets.v5i2.2125

Roff, K. A. (2018). Student Satisfaction and/or Dissatisfaction in Blended Learning Environments. *Frontiers in Education Technology*, *1*(2), 149. DOI: 10.22158/fet.v1n2p149

Ruokonen, I., & Ruismäki, H. (2016). E-Learning in Music: A Case Study of Learning Group Composing in a Blended Learning Environment. *Procedia: Social and Behavioral Sciences*, *217*, 109–115. DOI: 10.1016/j.sbspro.2016.02.039

Seman, S. A. A., Hashim, M. J., Roslin, R. M., & Ishar, N. I. M. (2019). Millennial learners' acceptance and satisfaction of blended learning environment. *Asian Journal of University Education, 15*(3), 129. Advance online publication. DOI: 10.24191/ajue.v15i3.7845

Shishakly, R., Almaiah, M. A., Al-Otaibi, S., Lutfi, A., Alrawad, M., & Almulhem, A. (2023). A New Technological Model on Investigating the Utilization of Mobile Learning Applications: Extending the TAM. *Multimodal Technologies and Interaction, 7*(9), 1–18. DOI: 10.3390/mti7090092

Xu, Z., Zhao, Y., Liew, J., Zhou, X., & Kogut, A. (2023). Synthesizing research evidence on self-regulated learning and academic achievement in online and blended learning environments: A scoping review. *Educational Research Review, 39*, 1–17. DOI: 10.1016/j.edurev.2023.100510

Yesmakhanova, Z., Zhussipova, G., Abjapparovna, K., Sumatokhin, S., Saparova, Z., & Assilbekova, B. (2022). Formation of digital competencies of future teachers of biology in a blended learning environment. *World Journal on Educational Technology: Current Issues, 14*(5), 1612–1621. DOI: 10.18844/wjet.v14i5.8123

Yuan, W. (2023). Perceptions of Chinese College Students towards Online Learning: Implications for Teaching in Blended Learning Environments. *The Educational Review, USA, 7*(2), 185–194. Advance online publication. DOI: 10.26855/er.2023.02.009

## KEY TERMS AND DEFINITIONS

**Accessibility:** The design of educational resources and environments that are usable by all students, including those with disabilities, ensuring equal opportunities for learning.

**Adaptive Learning:** A personalized learning experience that adjusts content and assessments based on individual learner performance and preferences, often powered by AI.

**Assessment:** The process of evaluating student learning through various methods, including formative (ongoing) and summative (final) assessments to gauge understanding and progress.

**Blended Learning:** An educational approach that combines traditional face-to-face instruction with online learning, allowing for flexibility and personalized experiences.

**Collaboration:** The act of working together with peers in learning activities, promoting social interaction and shared knowledge construction.

**Constructivism:** A learning theory emphasizing that knowledge is constructed through interaction with the environment and collaboration with others, encouraging active participation.

**Engagement:** The level of interest and participation a learner demonstrates in their educational experience, which can significantly impact learning outcomes.

**Gamification:** The use of game design elements in educational contexts to enhance motivation and engagement through rewards, challenges, and competition.

**Microlearning:** A strategy that delivers content in small, focused segments, allowing learners to absorb information more easily and efficiently.

**Professional Development:** Continuous education and training for educators to enhance their skills and competencies in teaching, particularly in integrating technology in learning environments.

# Chapter 6
# Mobile Learning and Digital Tools Effectiveness in Higher Education System in India

**Vijay Ganpat Dhamore**
https://orcid.org/0000-0002-5367-4725
*Ajeenkya D.Y. Patil University, India*

## ABSTRACT

*Fostering Cutting edge with Novelty in technology for knowledge sharing is a keystone in universal learning system. Mobile based Learning Management System (LMS) offers study material, online assessment, evaluation of quiz, result generation etc.to the higher education systems to all over world schooling organizations. Cloud Computing and Artificial intelligence skills are today involved in syllabus design of several curriculums in higher educational developments. This study will focus on higher education and the Digital tools or Mobile Apps used to teaching and learning monitoring systems in India for the Higher Education systems Purposive Sampling Method used for conducting the study. Primary Data collected through a Standard Questionnaire form and with the help of google form. Students from the Higher Education Students were respondents. Data analysis tools*

DOI: 10.4018/979-8-3693-8287-5.ch006

*like SPSS and SEM used to analyze the data. This study finally concludes for scope of LMS and Mobile Application and its effectiveness in Current Higher Education Systems in India.*

## INTRODUCTION

The adoption of Information and Communication Technology (ICT) has redefined pedagogical approaches in higher education. Mobile learning and digital platforms enable interactive, accessible, and flexible teaching methods that are increasingly essential in contemporary education. Tools such as smartboards, artificial intelligence–driven applications, and digital collaboration platforms have simplified complex learning processes and created new possibilities for engaging students. In India, mobile learning has been accelerated by widespread smartphone penetration and government-supported initiatives such as SWAYAM and DIKSHA. This chapter explores the effectiveness of mobile learning and digital tools in Indian higher education by examining their potential, challenges, and policy support.

Mobile learning (m-learning) represents more than a shift from traditional classroom instruction to digital alternatives; it is a transformation in the very philosophy of teaching and learning. The flexibility of mobile devices allows students to access learning materials anytime and anywhere, breaking the rigid boundaries of classroom walls and fixed schedules. This ubiquitous accessibility makes higher education more inclusive, particularly for learners from rural and semi-urban areas of India who might otherwise face geographical and infrastructural barriers. Moreover, the integration of multimedia—videos, podcasts, simulations, and interactive assessments—has enriched the learning experience, enabling learners to grasp complex concepts through visual and experiential means rather than solely through text.

The role of ICT in enhancing collaborative learning is particularly noteworthy. Online forums, peer-to-peer discussion boards, and group projects facilitated through cloud-based platforms encourage students to engage in active dialogue, problem-solving, and teamwork. This aligns with constructivist approaches to pedagogy, where learners construct knowledge through social interaction and shared experiences. In India, where classrooms are often characterized by large student-teacher ratios, digital tools provide a much-needed avenue for personalized feedback and interaction. Artificial intelligence–driven platforms can track learner progress, adapt content to individual needs, and recommend supplementary resources, thereby catering to diverse learning styles and abilities.

Government policies have been pivotal in creating a supportive ecosystem for digital learning. Initiatives such as SWAYAM (Study Webs of Active Learning for Young Aspiring Minds) and DIKSHA (Digital Infrastructure for Knowledge Sharing) provide free or affordable access to high-quality digital content developed by premier institutions. The National Education Policy (NEP) 2020 has further emphasized digital learning as a core driver of higher education reform, encouraging universities to adopt blended and online modes of instruction. These policy interventions highlight the recognition that ICT is not merely an add-on but a central component of future-ready higher education systems in India.

At the same time, mobile learning in India faces structural challenges that must be addressed to realize its full potential. Digital divide issues remain acute, with significant disparities in internet connectivity, bandwidth availability, and access to reliable devices across different socio-economic groups. Rural students, in particular, often face interruptions in learning due to weak connectivity and limited digital infrastructure. Moreover, the cost of data, though reduced in recent years, still imposes a financial burden on economically disadvantaged learners. While government initiatives provide free digital resources, the effectiveness of such programs depends on the ability of students to consistently access them.

Faculty readiness and pedagogical adaptation also play crucial roles in the successful integration of ICT. Many teachers in higher education institutions are still more comfortable with traditional face-to-face methods, and a lack of structured training often limits their ability to design effective online or blended learning experiences. Professional development programs focusing on digital pedagogy, instructional design, and technology adoption are necessary to bridge this gap. Furthermore, evaluation mechanisms for online learning outcomes need to be strengthened. While digital tools enable continuous assessment and analytics-based tracking, many institutions still rely heavily on summative examinations, which may not fully capture the skills and competencies developed through digital platforms.

The COVID-19 pandemic accelerated the adoption of mobile and digital learning, exposing both opportunities and shortcomings. On one hand, it demonstrated the resilience of ICT-enabled systems in ensuring continuity of education during unprecedented disruptions. Universities quickly shifted to platforms like Zoom, Google Meet, and Microsoft Teams, while students increasingly relied on smartphones for participation in online classes. On the other hand, the pandemic also exposed inequalities, as many students lacked the resources or environment to participate effectively. Post-pandemic, there is a growing recognition that hybrid learning models—combining online and offline methods—can offer the best of both worlds by ensuring flexibility while retaining the social and experiential benefits of face-to-face interaction.

Another critical dimension of ICT adoption in Indian higher education is the role of emerging technologies. Artificial intelligence, augmented reality (AR), and virtual reality (VR) are increasingly being explored to create immersive learning experiences. For example, AR/VR simulations can provide medical students with lifelike practice scenarios or engineering students with virtual labs. Gamification of learning—where elements of gaming such as points, levels, and challenges are integrated into educational content—can significantly boost motivation and engagement. These tools are not only reshaping content delivery but also preparing students for a digitally driven workforce, where technological literacy and adaptability are key competencies.

Equity and inclusivity remain central concerns in discussions of ICT in education. While digital learning expands opportunities for many, it risks leaving behind those without adequate access or support. Gender disparities, socio-economic differences, and disabilities all intersect with digital access in complex ways. For instance, in rural households where devices are shared, female students may have less time and autonomy to use digital resources. Similarly, learners with visual or auditory impairments may require specially designed accessible content, which is still limited in availability. Addressing these issues requires intentional policy design, inclusive content creation, and institutional support systems that ensure no learner is excluded.

In addition to accessibility, questions of digital literacy must be considered. For mobile learning to be effective, students and teachers must possess not only technical skills but also the critical ability to evaluate online information, manage digital identities, and engage responsibly in virtual environments. The rapid spread of misinformation and the potential for distractions in digital spaces underscore the need for embedding digital literacy as a core competency in higher education curricula.

From an institutional perspective, adopting ICT involves significant investment in infrastructure, content development, and faculty training. Universities must develop robust learning management systems (LMS), ensure cybersecurity, and continuously update digital resources to remain relevant. Public-private partnerships can play a key role here, bringing together the technological expertise of the corporate sector and the academic expertise of universities. Many Indian ed-tech companies, such as Byju's, Unacademy, and UpGrad, have already demonstrated the potential of digital platforms to scale learning across diverse audiences. Collaborations between such companies and higher education institutions can help bridge resource gaps and provide innovative solutions tailored to the Indian context.

Finally, the cultural shift brought about by ICT adoption in higher education must be acknowledged. The hierarchical, teacher-centered model of instruction traditionally dominant in India is gradually giving way to more learner-centered approaches, where students take greater responsibility for their own learning journeys. This shift requires a rethinking of assessment practices, curriculum design, and classroom dynamics. Mobile learning and digital tools encourage self-paced, autonomous learning while fostering creativity and critical thinking—skills essential for navigating a rapidly changing global landscape.

In conclusion, the adoption of ICT, mobile learning, and digital platforms has redefined the pedagogical landscape of Indian higher education. The promise of inclusivity, flexibility, and enhanced engagement makes digital learning an indispensable part of contemporary education. However, challenges related to infrastructure, equity, faculty readiness, and digital literacy must be systematically addressed to unlock its full potential. Government policies such as NEP 2020, SWAYAM, and DIKSHA provide a strong foundation, but their success depends on effective implementation and continuous adaptation to changing technological and social realities. As Indian higher education moves forward, a blended and inclusive digital ecosystem holds the key to preparing students for future challenges while democratizing access to quality learning opportunities.

## BACKGROUND

Mobile learning has evolved from supplementary digital support to a mainstream educational approach. Early initiatives in mobile-based education were limited in scope, but the convergence of high-speed connectivity, cloud platforms, and user-centered applications has expanded the role of digital tools. Research highlights the role of 5G, artificial intelligence, and immersive technologies such as Augmented Reality (AR) and Virtual Reality (VR) in shaping new learning ecosystems (Li & Wang, 2023; Song et al., 2020). In India, the National Education Policy (NEP) 2020 recognizes the transformative role of digital education in bridging equity gaps and democratizing access to higher education.

Mobile learning (m-learning) extends e-learning by leveraging handheld devices to support learning "anytime–anywhere," enabling continuity across settings and moments (Kukulska-Hulme & Traxler, 2019). Pedagogically, m-learning aligns with socio-constructivism—the idea that learners co-construct knowledge through interaction and scaffolding (Vygotsky, 1978)—and with conversational frameworks for iterative feedback and practice (Laurillard, 2013). In higher education, digital tools such as learning management systems (LMS), MOOCs, collaboration suites, AI-driven tutors, and AR/VR simulations enable richer modalities (Anderson, 2010; Radianti et al., 2020). Indian studies echo global findings that mobile devices foster immediacy, participation, and micro-learning, particularly when content is chunked and interactive (Gikas & Grant, 2013; Aithal & Aithal, 2016).

### Policy Landscape and National Platforms

India's policy architecture explicitly advances digital learning. The *National Education Policy (NEP) 2020* calls for mainstreaming online and blended modes, strengthening digital infrastructure, and building teacher capacity (Government of India, 2020). Parallelly, national platforms have scaled markedly. *SWAYAM* hosts interactive courses from school to post-graduate levels and supports credit transfer under UGC regulations, positioning MOOCs as a complement to campus learning (About SWAYAM, n.d.; SWAYAM, 2023). *DIKSHA*—initially school-focused—has driven content distribution, QR-linked textbooks, and teacher professional development at massive scale (Ministry of Education, 2020). UNESCO's financing profile reports *19,698+ courses, ~182 million enrolments, and ~145.7 million completions* on DIKSHA, illustrating India's reach in digital delivery (UNESCO, 2021). The *UGC Concept Note on Blended Mode* and subsequent guidance emphasize blended learning as impactful and encourage institutions to redesign curricula and assessment accordingly (UGC, 2021; AIU/CEMCA, 2024). AISHE reporting offers the annual statistical backdrop on infrastructure, enrolments, and the system's digital readiness (AISHE, 2024). Collectively, these frameworks institutionalize digital learning as a core, not peripheral, mode. (Government of India, 2020; About SWAYAM, n.d.; UNESCO, 2021; UGC, 2021; AIU/CEMCA, 2024; AISHE, 2024).

## Effectiveness: Access, Engagement, and Outcomes

*Access and flexibility.* M-learning improves access for commuting, working, and first-generation learners by decoupling study from place and schedule (Kukulska-Hulme & Traxler, 2019). Indian evidence from pandemic-period studies shows students valued recorded lectures, mobile-friendly notes, and asynchronous forums that allowed self-pacing and repetition—features linked to improved persistence (Dhawan, 2020; Muthuprasad et al., 2021). MOOC ecosystems (SWAYAM/NPTEL) provide structured syllabi, peer forums, auto-graded assessments, and proctored finals that can articulate into credits, improving perceived utility and legitimizing outcomes (About SWAYAM, n.d.; SWAYAM, 2023)

*Engagement and active learning.* Digital collaboration tools (discussion boards, shared docs, polls, breakout rooms) support dialogic learning and peer instruction (Anderson, 2010). Gamification elements (badges, points, leaderboards) can enhance motivation and time-on-task when aligned with clear learning outcomes (Dichev & Dicheva, 2017). Indian classroom studies report higher participation when instructors integrate formative quizzes and mobile-based polling into lectures, particularly in large-enrolment courses (Aithal & Aithal, 2016). Immersive tools (AR/VR) show positive effects on understanding complex spatial/procedural tasks in engineering, medicine, and lab-intensive disciplines (Radianti et al., 2020).

*Learning outcomes and analytics.* When m-learning is embedded within sound instructional design—clear objectives, scaffolded practice, timely feedback—studies report improved conceptual understanding and performance on applied tasks (Holmes et al., 2019; Siemens, 2014). Learning analytics derived from LMS/quiz data enable early alerts for at-risk students and iterative course improvements. Indian faculty highlight analytics dashboards as helpful for monitoring participation and tailoring remedial content, though institutional capacity to interpret data remains uneven (Muthuprasad et al., 2021).

## Inclusion, Equity, and the Digital Divide

Effectiveness is contingent on equitable access to *devices, bandwidth, and learning spaces*. Indian research documents sharp disparities by income, region, and gender in sustained access to smartphones, reliable data, and quiet study environments (Mehta & Shah, 2020; UNESCO, 2021). AISHE and ministry briefs acknowledge persistent infrastructure gaps, especially in rural and remote areas, that limit continuity and quality of online participation (AISHE, 2024; Behera/CIET brief on NEP's digital chapter). Student sharing of devices within households may constrain study time for women learners; accessible design (captions, transcripts, screen-reader-ready content) remains inconsistent (Goggin & Newell, 2020). Policy responses—community digital centres, subsidized data, open educational resources, and multilingual content—are necessary complements to platform expansion (Government of India, 2020; UNESCO, 2021).

## Faculty Readiness and Pedagogical Redesign

A recurring finding is that *technology alone does not yield impact*; instructor capacity and course redesign are decisive. Many Indian HEI faculty report comfort with face-to-face methods but less confidence in designing interactive, mobile-first learning experiences (Muthuprasad et al., 2021). NEP 2020 and UGC guidelines call for substantial professional development in digital pedagogy, assessment, and content authoring (Government of India, 2020; UGC, 2021). Effective practices include *chunked micro-lectures*, low-stakes quizzes for retrieval practice, structured discussion prompts, clear weekly pacing guides, and explicit norms for online collaboration (Anderson, 2010; Siemens, 2014). Institutions adopting "design teams" (faculty + instructional designers + media specialists) report smoother transitions and better student satisfaction.

## Assessment Integrity and Recognition

Mobile assessments enable frequent feedback but raise concerns about *integrity, proctoring, and alignment*. Proctored SWAYAM/NPTEL certifications and UGC credit-transfer rules help position MOOCs as credible parts of degree pathways (SWAYAM, 2023). Nevertheless, robust, *authentic assessment*—projects, case studies, viva-voce, and portfolios—remains key to curbing superficial learning and aligns better with employability skills (UGC, 2021). Blended models that reserve in-person time for demonstrations and capstones while using mobile tools for practice and formative checks show promise (AIU/CEMCA, 2024).

## Privacy, Safety, and Ethics

As platforms scale, *data governance* becomes integral to effectiveness and trust. India's DIKSHA ecosystem faced a reported data exposure incident that underscored the need for stronger security controls and privacy-by-design (Wired, 2023). Ethical analytics require transparency, minimal data collection, student consent, and bias monitoring in AI-driven recommendations (Holmes et al., 2019). Institutional policies and vendor contracts should specify data retention, usage limits, and grievance redressal; capacity-building for administrators and faculty on privacy norms is required

## Implementation Models Suited to Indian HEIs

The literature highlights several *fit-for-context models*:

- **Blended/flipped designs:** Short mobile-first pre-class videos, in-class problem-solving, and post-class retrieval practice improve conceptual change in quantitative disciplines (UGC, 2021; Dichev & Dicheva, 2017).
- **MOOC-enabled curricula:** SWAYAM/NPTEL for foundational/ theory components + in-house seminars, labs, and mentoring—backed by UGC credit transfer—expands elective choice without overburdening faculty (SWAYAM, 2023).
- **Micro-learning and WhatsApp/Teams channels:** Structured nudges, micro-quizzes, and peer help reduce procrastination and support commuter learners (Gikas & Grant, 2013).
- **Virtual labs/AR-VR modules:** For high-cost labs or safety-critical practice, immersive modules support skill rehearsal before physical labs (Radianti et al., 2020).
- **Faculty development "learning sprints":** Short cycles where instructors pilot one change (e.g., weekly low-stakes quiz) and review analytics to iterate (Siemens, 2014).

## Conditions for Effectiveness

Synthesizing across studies, *effectiveness* hinges on:

1. **Infrastructure and affordability**—reliable campus Wi-Fi, device-lending, low-bandwidth content formats, and offline-capable apps (Government of India, 2020; AISHE, 2024).
2. **Mobile-first instructional design**—brevity, interactivity, multimodality, and clear navigation (Anderson, 2010; Kukulska-Hulme & Traxler, 2019).

3. **Assessment redesign**—frequent formative checks on mobile, authentic summatives in blended settings (UGC, 2021).
4. **Faculty capability**—continuous CPD in digital pedagogy and analytics (Muthuprasad et al., 2021; Siemens, 2014).
5. **Equity measures**—accessible content, language support, and targeted aid for disadvantaged groups (UNESCO, 2021; Mehta & Shah, 2020).
6. **Governance**—privacy, cybersecurity, and interoperable systems; clear processes for credit transfer and recognition (SWAYAM/UGC).

## ISSUES, PROBLEMS, AND TRENDS IN MOBILE LEARNING

*Figure 1. Issues in ICT-enabled education*

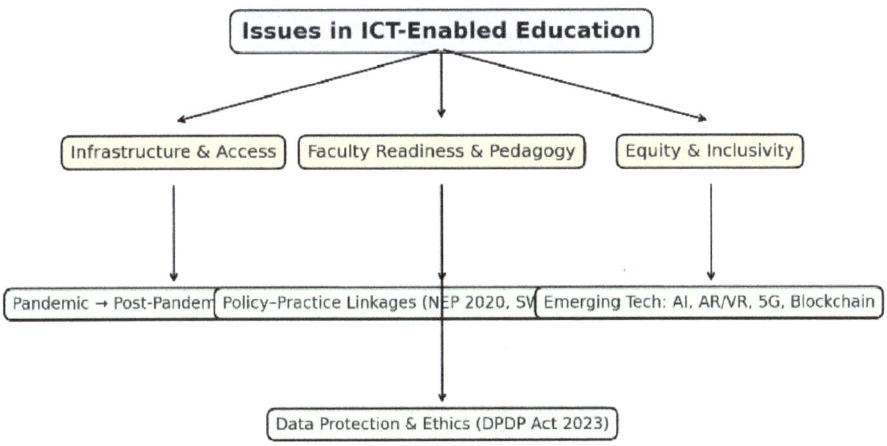


As Shown in Figure 1 Issues in ICT-Enabled Education the flowchart provides a organized impression of the issues, problems, and trends connected to ICT-enabled education in India. It highpoints the development from general issues to exact problems and developing trends, positioning them with policy, exercise, and future guidelines.

Mobile knowledge and digital tools have arisen as transformative forces in advanced education globally, yet their execution and amalgamation present a multifaceted set of tasks, predominantly in the Indian context. This section skeletons the key issues, problems, and trends that describe the existing state of mobile knowledge in higher education.

## 1. Issues

Issues mention to larger, systemic features that impact the application and usefulness of mobile learning.

a. Digital Divide and Unfairness
b. Organization Boundaries
c. Faculty Numerical Readiness
d. Policy Application Gaps
    - o Despite initiatives like SWAYAM and NEP 2020 encouraging digital education, gaps persist in institutional acceptance, funding apportionment, and arrangement with curriculum objects.

## 2. Problems

Problems are the practical, operational encounters that organizations, instructors, and students face during mobile learning adoption.

a. Affordability and Cost Limitations
b. Usability and User Understanding
c. Content Significance and Localization
d. Data Confidentiality and Safety Concerns

## 3. Trends

Trends highpoint the way in which mobile learning adoption and novelty are moving.

a. Integration of Evolving Technologies
b. Post-Pandemic Digital Quickening
c. Shift Toward Microlearning and On-Demand Learning

d. Policy Provision for Digital Learning
e. Increased Prominence on Learner-Centric Design

While mobile learning offers transformative opportunities, several challenges persist in India. Systemic issues include the digital divide, inadequate infrastructure, and faculty readiness. Operational problems such as affordability of devices, usability of platforms, and concerns over data privacy remain barriers to adoption. At the same time, trends such as AI-enabled personalized learning, microlearning modules, and post-pandemic hybrid models are reshaping the educational landscape. Addressing these gaps through targeted policy interventions and capacity building is essential to unlock the full potential of mobile learning in higher education.

## TECHNOLOGICAL INFRASTRUCTURE AND ADVANCEMENTS

A reliable theme in the literature is the role of progressive announcement technologies, chiefly 5G and edge computation, in enabling richer and more accessible learning involvements. Li and Wang (2023) established that artificial intelligence and edge-computing knowledges can digitalize and envisage development materials, safeguarding continuous user-friendliness both online and offline. Their study highlighted that 5G wireless announcement significantly enhances teaching quality through more effective valuation and response mechanisms. Cheng et al. (2021) prolonged this conversation by scrutinizing classroom teaching excellence evaluation systems, concluding that the integration of 5G with data-driven teaching valuations offers smarter, more precise, and computerized evaluation processes. Song et al. (2020) enclosed 5G as a foundation of a broader digital uprising in education, enabling consistent learning environments and a miscellaneous range of intellectual applications. While these studies underscore technological benefits, they offer limited exploration of implementation inequalities in resource-constrained contexts, an area that remains underexplored in Indian higher education.

## DIGITAL EXPERIENCE PLATFORMS AND ECOSYSTEM INTEGRATION

The growth of Digital Experience Platforms (DXPs) imitates a shift towards holistic digital ecosystems. Qudus and Ibra (2024) argued that DXPs unify content, commerce, and engagement into all-in-one learner experiences, emphasizing adaptability and user-centered design. The definition of DXPs highlights their addition of content management, analytics, and collaboration tools, enabling consistent experiences across multiple networks. Although promising, these stages require noteworthy institutional investment, raising questions about scalability and equity in diverse higher education settings.

## EVOLUTION OF MOBILE LEARNING

The history of mobile learning (M-learning) can be traced to early creativities such as Grant's (1993) Wireless Coyote project, which discovered the use of mobile computers for field-based learning. Brown (2005) sited mobile learning as an postponement of e-learning, enabling greater accessibility. Over time, mobile learning has progressed from simple portability (Naismith et al., 2004; Andronico et al., 2014) to integration with emerging technologies such as AI, IoT, and VR (Sharma, 2019). This progression exemplifies a paradigm shift—from mobile learning as a additional option to a typical educational modality—yet adoption paths vary significantly between foundations and provinces.

## ADOPTION OF MOBILE LEARNING TECHNOLOGIES

The acceptance process replicates both individual excellent and institutional willingness. Rogers (2003) described adoption as the procedure by which students choose to participate mobile learning into their enlightening observes. Musa (2024) provided empirical evidence from 680 higher education students, endorsing targeted efforts to bridge the digital divide, particularly in rural areas, through supports and locally industrialized knowledge tools. Notwithstanding such understandings, the literature often superintends cultural and educational factors manipulating implementation in Indian higher schooling contexts, on behalf of a key research gap.

## USE AND USABILITY CONSIDERATIONS

Actual mobile learning be contingent not only on device possession but also on readiness and capability to use technology for knowledge purposes (Hao et al., 2017). Ratna (2020) prolonged the notion of "user-friendliness" into broader usability, highlighting the need for intuitive interfaces, logical steering, and helpful media. Economides and Nikolaou (2008) emphasized design dimensions—including physical features and interface quality that effect continued usage. However, the literature delivers limited cross-analysis of how serviceability anxieties align with educational efficiency.

## FACTORS INFLUENCING USE IN HIGHER EDUCATION

Several studies identify recurrent factors of mobile learning adoption and effectiveness. Shuja et al. (2019), Crompton (2013), Guo (2014), and Aremu & Adeoluwa (2021) cited flexibility, convenience, cooperative learning, personalization, and transferability as key enablers. These influences align with broader acceptance models such as the Technology Acceptance Model (TAM) and Unified Theory of Acceptance and Use of Technology (UTAUT), though the educations rarely frame their findings explicitly within such theoretical constructs. A critical remark is the lack of longitudinal educations measuring how these factors evolve with technological and educational novelties.

## POLICY AND DIGITAL INITIATIVES IN INDIA

The scenery of higher education in India has been experiencing a noteworthy alteration, ambitious by national guidelines and large-scale government creativities that highlight the addition of skill into teaching and learning. Central to this reform is the national education policy (NEP) 2020, which imagines a holistic, multidisciplinary, and technology-enabled education system. NEP 2020 places digital learning at the core of instructive reform by emphasizing equity, accessibility, and lifelong learning opportunities. Its vision aligns with global trends in digital pedagogy while contextualizing them to address India's diverse socio-economic and linguistic realities.

The NEP 2020 explicitly indorses the creation of a national educational technology forum (NETF) to foster the exchange of ideas, enable capacity construction, and guide the integration of emerging technologies such as artificial intelligence, adaptive learning platforms, and online content delivery.

In line with these objectives, numerous government-supported e-learning platforms have been developed and scaled in recent years. Among them, Swayam (study webs of active-learning for young aspiring minds) has emerged as India's flagship massive open online course (MOOC) initiative.

Complementing Swayam is the diksha (digital infrastructure for knowledge sharing) platform, which caters specifically to teachers and school-level learners. Developed by the ministry of education, diksha serves as a national repository of curriculum-aligned content, interactive resources, and teacher professional development modules.

India's digital education initiatives collectively represent a strategic effort to build a strong technology-enabled learning ecosystem that bridges access, quality, and equity gaps. At the school level, the PM e-VIDYA program launched in 2020 consolidated multiple platforms, including DIKSHA, Swayam Prabha television channels, and radio-based content, ensuring instructional continuity during the COVID-19 pandemic. For higher education, resources such as NPTEL and e-PG Pathshala have played a central role in expanding discipline-specific knowledge through MOOCs and comprehensive postgraduate e-resources, thereby fostering self-directed and lifelong learning. These efforts, alongside NEP 2020's vision, highlight India's alignment with global digital learning movements while tailoring reforms to local socio-economic contexts. Importantly, the integration of such platforms has been accompanied by pedagogical innovations. Approaches like blended learning—which combines face-to-face instruction with online modules—are widely practiced in institutions such as the IITs and IIMs, while flipped classrooms have gained traction in medical and engineering education, enabling students to engage with learning materials independently before applying knowledge in interactive classroom settings. Furthermore, the Community of Inquiry (COI) framework offers a theoretical basis for evaluating online learning quality, emphasizing cognitive, social, and teaching presence. Applied within Indian MOOCs such as SWAYAM, the COI model illustrates how meaningful interaction and teacher facilitation can enhance learning outcomes. Together, these models mark a pedagogical shift from teacher-centered methods to learner-focused ecosystems, reinforcing NEP 2020's emphasis on flexibility, critical thinking, and creativity in higher education.

## BARRIERS TO TECHNOLOGY-ENABLED LEARNING IN INDIA:

These barriers reflect systemic inequities that can significantly influence digital adoption among learners and faculty in the study.

*Table 1. Barriers to technology-enabled learning in India*

| Barrier | Description | Data highlights |
|---|---|---|
| Connectivity | Access vs. Actual use | 95% villages connected, rural internet usage at 45 per 100, vs 110 in urban |
| Digital divide | Urban–rural penetration disparity | Rural: 40–42%, urban: ~58% internet penetration |
| Device affordability | High cost, limited ownership | Only 4.2% rural vs 21.6% urban households own computers |
| Digital literacy | Knowledge and skills gap | Rural literacy ~25%, urban ~61%; only 21% st digitally literate |
| Infrastructure limits | Terrain and power constraints | Fiber deployment challenges, unreliable electricity in remote areas |

As Shown in Table 1 Barriers to Technology-Enabled Learning in India its description presented in a summary as well as illustrative way. Evidence from India underscores the importance of strengthening *facilitating conditions*—such as device availability, reliable internet access, and stable platforms—while enhancing *performance expectancy* through interactive, high-quality course design and embedding blended learning practices that foster both social and teaching presence. These strategies are consistent with TAM and UTAUT adoption models as well as diffusion-through-reinvention, and they address the challenges observed in platforms like SWAYAM and NPTEL. The COVID-19 pandemic served as a turning point, forcing institutions to transition rapidly to online education between 2020 and 2022. During this period, national platforms such as DIKSHA and PM e-VIDYA witnessed unprecedented use, while SWAYAM reported sharp enrolment surges. Increased reliance on digital resources coincided with a rise in mobile data consumption, which remains significantly higher post-pandemic, reflecting a long-term shift in study and media habits. Policy reforms also formalized online learning as part of the credit framework, with the UGC permitting up to 40% of coursework to be delivered via SWAYAM, and subsequent initiatives institutionalizing blended and flexible pathways. Since 2022, many universities have stabilized into hybrid formats that combine MOOCs with on-campus labs and assessments, revealing new challenges centered on pedagogy, interaction, and completion rates. At the same time, equity concerns remain prominent, as rural learners still face gaps in internet access, affordability, and digital literacy, making multimodal delivery—including TV, radio, and low-bandwidth mobile solutions—essential. Within this study, effectiveness is assessed across four dimensions: (1) learning outcomes, measured by alignment of student performance with course objectives and CO–PO attainment; (2) engagement levels, tracked through platform usage and participation metrics; (3) satisfaction, gauged via structured surveys on usability, relevance, and support; and (4) academic performance, evaluated through examination results, progression, and triangulated feedback. This multidimensional evaluation provides a rigorous lens for assessing not only immediate learning outcomes but also long-term readiness for blended and digital-first higher education models in India.

## GAPS AND RESEARCH OPPORTUNITIES

Across these thematic areas, the literature tends to emphasis seriously on technological competences while if less critical appointment with academic integration, faculty willingness, and institutional policy outlines. There is also a paucity of context-specific investigate for Indian higher education, particularly in relation to equity of access, regional differences, and impact valuation over time.

Overall, the literature founds a strong foundation for sympathetic mobile learning's potential, but it also reveals clear gaps in critical amalgamation, theoretical framing, and related pertinence. By addressing these limitations, the present study aims to contribute a more nuanced, analytically grounded sympathetic of mobile learning acceptance and use in Indian higher education. Despite progress, gaps persist. Rigorous impact evaluations remain limited; many reports rely on satisfaction or usage proxies rather than learning gains and labour-market outcomes. There is scope for randomized or quasi-experimental studies on specific mobile interventions in Indian heis, with attention to heterogeneity (gender, rurality, first-generation status). Accessibility research should expand beyond captions to universal design at scale in indian languages. The ethics of ai tutors and algorithmic transparency in recommendation engines warrant scrutiny in Indian contexts. Finally, faculty workload models and incentives must evolve so redesign labour is recognized in appraisal and promotion.

### Aims and Objectives

The aim of the study is to find the correlation between adoption, and use of mobile learning technologies with traditional learning methods among Nigerian university students

1. To determine the factors that influence the adoption, and use of mobile learning technologies among students.
2. To carry out a comparative analysis of mobile learning technologies and traditional learning methods among students.

## Hypotheses

The following null hypotheses tested in this research:

**H01:**There is no significant relationship between Preference of Mobile Technology and the Level of Adoption for Learning Management System among students in at ADYPU,Pune India.
**H02:**There is no significant relationship between Level of adoption of mobile learning technologies and Experience of Mobile Leaning Technology among students in at ADYPU,Pune India.

## METHODOLOGY

The study employed a quantitative research design using purposive sampling. Data were collected from 120 students of Ajeenkya DY Patil University, Pune, through a structured questionnaire. The questionnaire was based on a 5-point Likert scale to capture student perceptions regarding mobile learning tools. Statistical analyses were conducted using SPSS and Structural Equation Modeling (SEM) to examine relationships among key variables, including preference, level of adoption, and experience of mobile learning technologies.

*Figure 2. Conceptual research model*

As Shown in Figure 2 Conceptual Research Model describes the Variables and its relationship.

## Target Population

The sum of items or objects that the researcher is interested in referred to as the population. It can also refer to the overall number of people in a particular field of study. As a result, the population of this study included all students at the Ajeenkya DY Patil University, Pune, India total of 120 students, of whom 35 (29%) were undergraduate students and 85 (71%) were postgraduate students.

## Sampling Techniques

Purposive sampling was used in this study; this type of sampling approach was used on a homogeneous population.

## Sample Size

The data for this study were gathered by means of a survey questionnaire administered to 120. Undergraduate and Postgraduate students during the 2024/2025 session. The survey instructed students to provide feedback about their experiences with the e-learning system.

A sample is a smaller group of subjects drawn from the population in which a given study was conducted for a purpose of drawing conclusions about the population targeted. For example, [19] argued that the result from the sample could be used to generalize about the entire population as long as it is truly represented.

## Instruments for Data Collection

A questionnaire was constructed using a 5-point scale and administered to collect the data for this research: 1-Strongly Agree, 2- Agree, 3-Neutral, 4-Disagree, and 5-Strongly Disagree. This survey instrument comprises of two parts, one for each demographic section and the main instruments (including factors that affect LMS learning technologies and knowledge and skill on the mobile technologies). This study investigates two Mobile learning categories which are the technology factors affecting the use of Mobile Learning and secondly the knowledge and skill on using the mobile learning technologies.

## RESULTS AND DISCUSSION

Statistics and correlation instrument, was coded, and analyzed using Statistical Package for Social Sciences (SPSS).

*Table 2. Scale statistics*

| **Scale Statistics** | | | |
|---|---|---|---|
| **Mean** | **Variance** | **Std. Deviation** | **N of Items** |
| 57.25 | 56.119 | 7.491 | 16 |

As Shown in Table 2 Scale Statistics Cronbach's Alpha Value is 0.862 is which is near to 0.90. So all the Questions in the Questionnaire is Reliable and Valid.

*Table 3. Reliability statistics*

| **Reliability Statistics** | | |
|---|---|---|
| **Cronbach's Alpha** | **Cronbach's Alpha Based on Standardized Items** | **N of Items** |
| .876 | .862 | 16 |

As shown in Table 3 Reliability Statistics the values Cronbach Alpha if more than 07

*Figure 3. CFA path diagram*

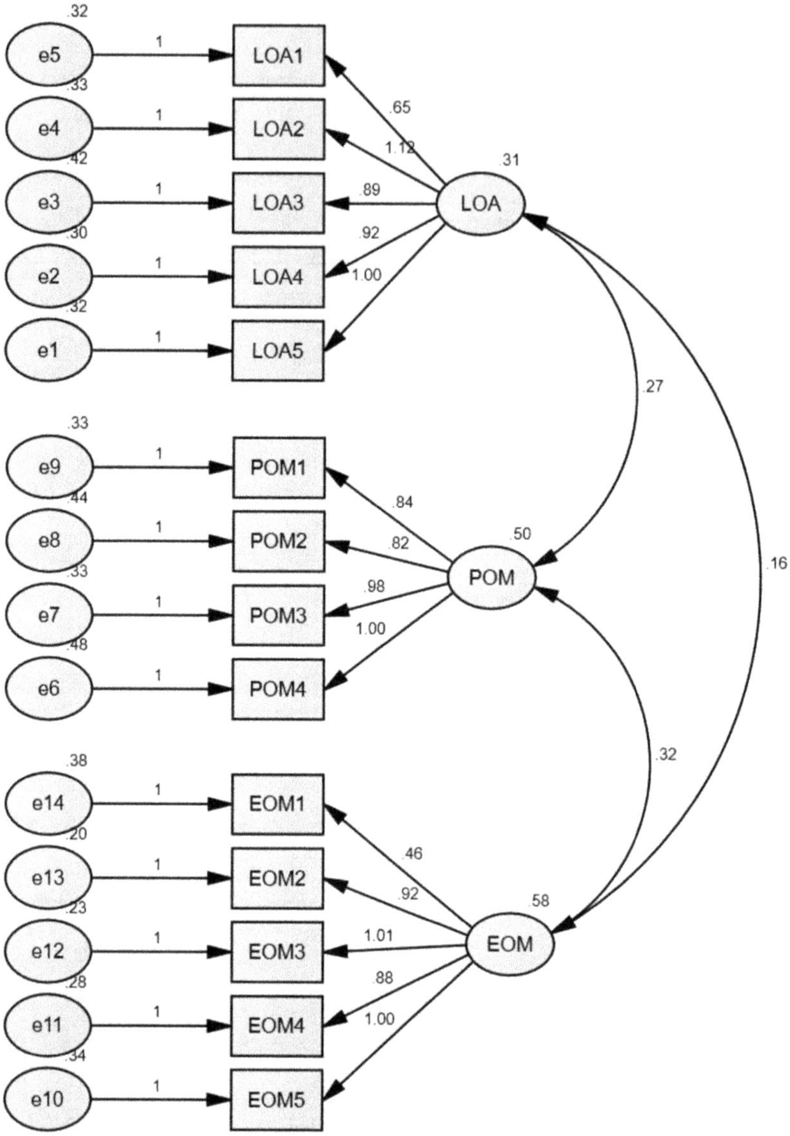


As Shown in Figure 3 CFA Path Diagram Clearly shows the Correlation Fit among the Variables under study.

*Figure 4. Structural model*

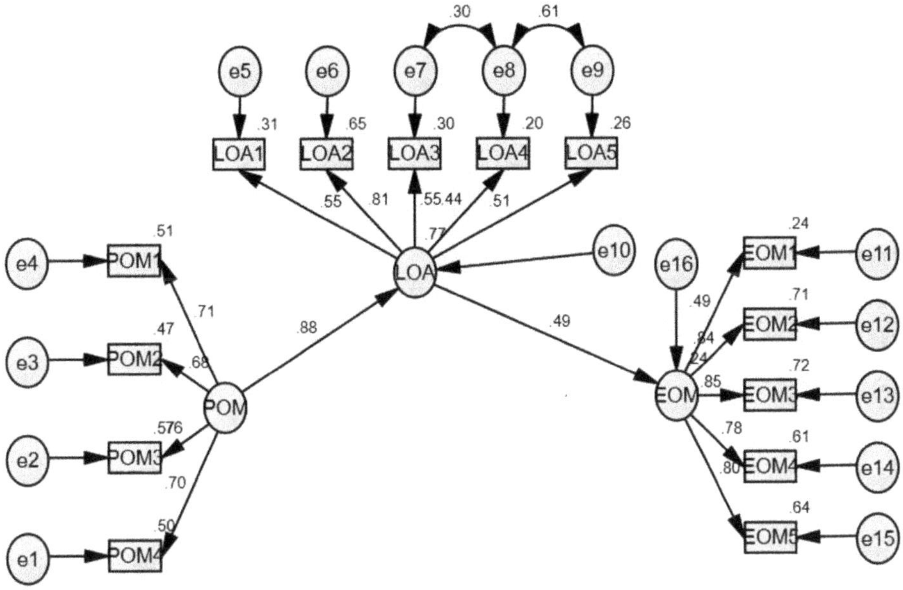


As Shown in Figure 4 Structural Model Clearly shows the Causal Relationship among the Variables under study.

## Standardized Regression Weights: (Group Number 1 - Default Model)

*Table 4. Standardized regression weights*

| | | | Estimate |
|---|---|---|---|
| **LOA** | **<---** | **POM** | **.877** |
| **EOM** | **<---** | **LOA** | .490 |
| **POM4** | **<---** | **POM** | .705 |
| **POM3** | **<---** | **POM** | .757 |
| **POM2** | **<---** | **POM** | .683 |
| **POM1** | **<---** | **POM** | .711 |
| **LOA1** | **<---** | **LOA** | .555 |

continued on following page

*Table 4. Continued*

| | | | Estimate |
|---|---|---|---|
| **LOA** | <--- | **POM** | **.877** |
| **LOA2** | <--- | **LOA** | .805 |
| **LOA3** | <--- | **LOA** | .549 |
| **LOA4** | <--- | **LOA** | .442 |
| **LOA5** | <--- | **LOA** | .512 |
| **EOM1** | <--- | **EOM** | .490 |
| **EOM2** | <--- | **EOM** | .844 |
| **EOM3** | <--- | **EOM** | .851 |
| **EOM4** | <--- | **EOM** | .781 |
| **EOM5** | <--- | **EOM** | .799 |

As Shown in Table 4 Standardized Recession weights are mentioned and all are above 0.5

## CMIN

*Table 5. Model fit statistics*

| Model | NPAR | CMIN | DF | P | CMIN/DF |
|---|---|---|---|---|---|
| Default model | 32 | 328.149 | 73 | .000 | 4.495 |
| Saturated model | 105 | .000 | 0 | | |
| Independence model | 14 | 1196.354 | 91 | .000 | 13.147 |

## Model Fit Summary

As Shown in Table 5 Model Fit Statistics CMIN and P Values are Showed and all are within the acceptable limits.

*Table 6. RMR, GFI*

| Model | RMR | GFI | AGFI | PGFI |
|---|---|---|---|---|
| Default model | .075 | .783 | .687 | .544 |
| Saturated model | .000 | 1.000 | | |
| Independence model | .260 | .338 | .236 | .293 |

As Shown in Table 6 RMR, GFI Values are Showed and all are within the acceptable limits.

*Table 7. Baseline comparisons*

| Model | NFI Delta1 | RFI rho1 | IFI Delta2 | TLI rho2 | CFI |
|---|---|---|---|---|---|
| Default model | .726 | .658 | .773 | .712 | .769 |
| Saturated model | 1.000 | | 1.000 | | 1.000 |
| Independence model | .000 | .000 | .000 | .000 | .000 |

As Shown in Table 7 Baseline Comparisons Values are Showed and all are within the acceptable limits.

## STRUCTURAL MODEL ASSESSMENT

A Structural Equation Model generated through AMOS Was used to test the relationships. A Good fitting model is accepted if the value of the CMIN/df is < 5, the goodness of fit (GFI) indices, (Hair et al, 2010); the Tucker and Lewis Index, 1973 (TLI); the confirmatory fit index((CFI) (Bentler,1990) is > 0.90 (Hair et al, 2010).In addition, an adequate fitting model was accepted if the AMOS computed model value of the Standardized root mean square residual RMR< 0.05 and the root mean square error approximation (RMSEA) is between 0.05 and 0.08 (Hair et al, 2010).The fit indices for the model shown in Table No.04.fell within the acceptable range: CMIN/df =4.495,the goodness of fit (GFI)=0.783,TLI=0.712,C-FI=0.769 and RMSEA-0.156.The Squared multiple correlation was 0.769 for Level of Adoption. The study assessed the impact of Preference of mobile on Level of Adoption is 87%. The impact of Level of Adoption on Experience of Mobile technology is 43% was Positive. Model fit indices and Hypotheses results are presented in Table 6.

*Table 8. Structural model assessment summary*

| Hypothesized Relationship | Standardized Estimates | t-value | p-value | Decision |
|---|---|---|---|---|
| POM ->LOA | 0.877 | 0.086 | *** | Accepted |
| LOA ->EOM | 0.490 | 0.124 | *** | Accepted |

CMIN/df =4.495,the goodness of fit (GFI)=0.783,TLI=0.712,CFI=0.769 and RMSEA-0.156

b=Standardized Regression Weight.

As Shown in Table 8 Structural Model Assessment Summary is well described. The findings indicate that students perceive mobile learning tools as reliable and effective, with a Cronbach's alpha of 0.862 demonstrating strong reliability of the survey instrument. The structural model revealed a strong relationship between preference for mobile technologies and level of adoption ($\beta = 0.877$, $p < .001$), as well as a positive relationship between adoption and experience of mobile technologies ($\beta = 0.490$, $p < .001$). These findings align with the Technology Acceptance Model (Davis, 1989) and Unified Theory of Acceptance and Use of Technology (Venkatesh et al., 2003), emphasizing perceived usefulness, ease of use, and facilitating conditions as key determinants of technology adoption.

## FINDINGS

The study highlights several insights: (1) students demonstrate a positive attitude toward mobile learning, (2) policy initiatives such as NEP 2020, SWAYAM, and DIKSHA provide strong support but require improved implementation, and (3) barriers such as infrastructure limitations and faculty digital readiness continue to restrict effectiveness. Learner inclusivity, equitable access, and data privacy remain pressing concerns that require systematic attention.

## SOLUTIONS AND RECOMMENDATIONS

To strengthen mobile learning in Indian higher education, this chapter recommends: (a) continuous professional development for faculty to enhance digital literacy, (b) targeted investments in infrastructure to reduce connectivity and access gaps, (c) affordable device and data packages to address equity issues, (d) robust data protection policies aligned with the Digital Personal Data Protection Act, 2023, and (e) curriculum design that integrates blended and learner-centered pedagogical models.

## FUTURE RESEARCH DIRECTIONS

Future research should explore the integration of emerging technologies such as AI, AR/VR, and 5G-enabled IoT in mobile learning. Longitudinal studies assessing the sustainability of blended learning models in Indian higher education are also essential. Research into microlearning, gamification, and blockchain-based credentialing can further expand the understanding of mobile learning effectiveness. Emerging technologies on the unexplored areas like Artificial Intelligence and Adaptive Learning, Augmented and Virtual Reality (AR/VR), 5G and IoT Integration, Gamification and Microlearning, Blockchain and Credentialing to be essential in nearest future.

## CONCLUSION

Mobile learning and digital tools are reshaping higher education in India by expanding access, flexibility, and learner engagement. However, systemic inequities and implementation challenges must be addressed to realize their full potential. By aligning technological adoption with inclusive pedagogical strategies and robust policy frameworks, India can build a more equitable and sustainable higher education ecosystem. The future of mobile learning lies in its capacity to integrate emerging technologies while ensuring accessibility and inclusivity for all learners. Forthcoming reiterations of this work should integrate Indian-specific evidence, highlight effective educational practices, involve with policy frameworks, and consider ethical concerns like data protection. Such rebalancing will safeguard the chapter speaks to both academic and practitioner audiences, enhancing its policy relevance and practical utility. By positioning mobile learning within a broader ecosystem of pedagogy, policy, and equity, the chapter can make a more all-inclusive and impactful contribution to evolving higher education in India.

## REFERENCES


About, S. W. A. Y. A. M. (n.d.). *Swayam Central*. https://swayam.gov.in/about

Aithal, P. S., & Aithal, S. (2016). Impact of online education on higher education system. *International Journal of Engineering Research and Modern Education, 1*(1), 225–235.

Anderson, T. (2010). *The theory and practice of online learning* (2nd ed.). Athabasca University Press.

Andronico, P., Minutoli, S., & Kuruoglu, E. E. (2014). Understanding elderly needs for designing a digitally extended environment via tablets. In Paternò, F., Santoro, C., & Tarantino, L. (Eds.), *Human-computer interaction* (pp. 284–287). Springer., DOI: 10.1007/978-3-319-07854-0_33

Aremu, B. V., & Adeoluwa, O. V. (2021). M-learning: A nexus for adult learners' motivation and readiness to learn in federal universities at Southwest, Nigeria. *Journal of Digital Educational Technology, 2*(1), ep2201. Advance online publication. DOI: 10.21601/jdet/11361

Brown, T. H. (2005). Towards a model for m-learning in Africa. *International Journal on E-Learning, 4*(3), 299–315.

Cheng, Q., Li, B., & Zhou, Y. (2021). Research on evaluation system of classroom teaching quality in colleges and universities based on 5G environment. In *Proceedings of the 2021 International Conference on Control and Intelligent Robotics* (pp. 74–85). ACM. DOI: 10.1145/3473714.3473728

Crompton, H. (2013). *Handbook of mobile learning*. Routledge.

Davis, F. D. (1989). Perceived usefulness, perceived ease of use, and user acceptance of information technology. *Management Information Systems Quarterly, 13*(3), 319–340. DOI: 10.2307/249008

Dichev, C., & Dicheva, D. (2017). Gamifying education: What is known and what remains uncertain. *International Journal of Educational Technology in Higher Education, 14*(9), 1–21.

Economides, A. A., & Nikolaou, N. (2008). Evaluation of handheld devices for mobile learning. *International Journal of Engineering Education, 24*(1), 3–13.

Gikas, J., & Grant, M. M. (2013). Mobile computing devices in higher education: Student perspectives. *The Internet and Higher Education, 19*, 18–26. DOI: 10.1016/j.iheduc.2013.06.002

Goggin, G., & Newell, C. (2020). *Digital disability: The social construction of disability in new media.* Routledge.

Government of India. (2020). *National Education Policy 2020.* Ministry of Education. https://www.education.gov.in/nep2020

Grant, W. C. (1993). Wireless Coyote: A computer-supported field trip. *Communications of the ACM, 36*(5), 57–59. DOI: 10.1145/155049.155062

Hao, S., Dennen, V. P., & Mei, L. (2017). Influential factors for mobile learning acceptance among Chinese users. *Educational Technology Research and Development, 65*(1), 101–123. DOI: 10.1007/s11423-016-9465-2

Holmes, W., Bialik, M., & Fadel, C. (2019). *Artificial intelligence in education: Promises and implications for teaching and learning.* Center for Curriculum Redesign.

Kukulska-Hulme, A., & Traxler, J. (2019). *Mobile learning: The next generation.* Routledge.

Laurillard, D. (2013). *Rethinking university teaching: A conversational framework for the effective use of learning technologies* (2nd ed.). Routledge. DOI: 10.4324/9781315012940

Li, F., & Wang, C. (2023). Artificial intelligence and edge computing for teaching quality evaluation based on 5G-enabled wireless communication technology. *Journal of Cloud Computing (Heidelberg, Germany), 12*(45), 45. Advance online publication. DOI: 10.1186/s13677-023-00418-6

Mehta, S., & Shah, H. (2020). Digital divide and online education in India. *Journal of Educational Planning and Administration, 34*(2), 123–138.

Ministry of Education. (2020). *DIKSHA: Digital Infrastructure for Knowledge Sharing.* Government of India.

Ministry of Education. (2023). *All India Survey on Higher Education (AISHE) 2022–23*. Government of India.

Muthuprasad, T., Aiswarya, S., Aditya, K. S., & Jha, G. K. (2021). Students' perception and preference for online education in India during COVID-19 pandemic. *Social Sciences & Humanities Open*, *3*(1), 100101. DOI: 10.1016/j.ssaho.2020.100101 PMID: 34173507

. Naismith, L., Lonsdale, P., Vavoula, G., & Sharples, M. (2004). *Literature review in mobile technologies and learning* (Futurelab Series Report 11). Futurelab.

Radianti, J., Majchrzak, T. A., Fromm, J., & Wohlgenannt, I. (2020). A systematic review of immersive virtual reality applications for higher education. *Computers & Education*, *147*, 103778. DOI: 10.1016/j.compedu.2019.103778

Rogers, E. M. (2003). *Diffusion of innovations* (5th ed.). Free Press.

Sharma, P. (2019). Digital revolution of Education 4.0. *International Journal of Engineering and Advanced Technology*, *9*(2), 3558–3564. DOI: 10.35940/ijeat.A1293.129219

Shuja, A., Qureshi, I. A., Schaeffer, D. M., & Zareen, M. (2019). Effect of m-learning on students' academic performance mediated by facilitation discourse and flexibility. *Knowledge Management & E-Learning*, *11*(2), 158–200. DOI: 10.34105/j.kmel.2019.11.009

Siemens, G. (2014). Learning analytics: The emergence of a discipline. *The American Behavioral Scientist*, *57*(10), 1380–1400. DOI: 10.1177/0002764213498851

Song, H., Liu, Y., Li, K., Ding, L., Sun, X., & Zhang, X. (2020). Reform and future prospects in education based on 5G technology. In Amornpitoksuk, P. (Ed.), *Proceedings of the 3rd International Conference on Education Technology and Information System (ETIS 2020)* (pp. 136–142). Clausius Scientific Press.

SWAYAM. (2023). *SWAYAM annual report 2022–23*. Ministry of Education, Government of India.

UGC. (2021). *Guidelines on blended learning and evaluation reforms.* University Grants Commission.

UNESCO. (2021). *Reimagining our futures together: A new social contract for education.* UNESCO Publishing. https://unesdoc.unesco.org/ark:/48223/pf0000379707

Venkatesh, V., Morris, M. G., Davis, G. B., & Davis, F. D. (2003). User acceptance of information technology: Toward a unified view. *Management Information Systems Quarterly*, *27*(3), 425–478. DOI: 10.2307/30036540

Vygotsky, L. S. (1978). *Mind in society: The development of higher psychological processes.* Harvard University Press.

Wired. (2023, January 23). A major app flaw exposed the data of millions of Indian students. *Wired.* https://www.wired.com/story/india-diksha-app-data-leak

## KEY TERMS AND DEFINITIONS

**ADYPU:** Ajeenkya DY Patil University in this study abbreviated as ADYPU,Pune,India

**Acceptance of Mobile Learning Technologies:** the willingness and readiness of students to adopt and use mobile learning technologies or platforms for educational purposes (Dillon & Morris 1996).

**Adoption of Mobile Learning Technologies:** the process by which students choose to integrate and use mobile learning technologies or platforms most especially for their learning experiences.

**Mobile Learning:** is the use of mobile technologies to access learning resources from anywhere at any time.

**Mobile Learning Technologies:** the use of mobile learning devices such as smart phones, tablets. Personal digital assistants (pdas) to facilitate and enhance learning experience.

**Traditional Learning Methods:** the use of instructional methods like dictation, drill and practice, and expository learning in a teacher led classroom settings.

**Use of Mobile Learning Technologies:** the application of mobile learning technologies for learning purposes.

Chapter 7

# Next Gen Education Elevating Higher Learning With Smart and Mobile Applications

**Surenthran David P.**
https://orcid.org/0009-0001-7665-2406
*Xavier Institute of Management and Entrepreneurship, Chennai, India*

**Daniel Thangadurai Hepzibah Vinsyah Jeyaseeli**
https://orcid.org/0009-0002-7218-7444
*Francis Xavier Engineering College, Trinelveli, India*

**Joel Jebadurai Devapitchai**
https://orcid.org/0000-0002-6947-8497
*St. Joseph's College of Engineering, OMR, Chennai, India*

**M. Nanda kumar**
*Gojan School of Business and Technology, India*

**M. M. Gayathri**
https://orcid.org/0009-0001-8625-8023
*Adhiparasakthi College of Engineering, India*

**S. Vijayalakshmi**
https://orcid.org/0009-0004-7558-2479
*Jaya College of Arts and Science, India*

DOI: 10.4018/979-8-3693-8287-5.ch007

## ABSTRACT

*In the technological era, smart and mobile applications are playing a crucial role in enhancing the learning experience of students to the next level. To understand its impact on higher education, this chapter explores the application of Augmented Reality (AR) and Virtual Reality (VR) to build immersive learning environments, as well as the integration of Artificial Intelligence (AI) to customize educational content to each student's needs. It also addresses the challenges associated with mobile learning, such as the digital divide, security concerns, and the need for effective pedagogy. Furthermore, the chapter discusses future trends, including the rise of micro learning, gamification, and the impact of 5G technology. By examining case studies and success stories, the chapter provides a comprehensive overview of how smart and mobile applications are reshaping higher education, ultimately seeking to establish a more welcoming learning environment.*

## INTRODUCTION

The introduction of mobile technology and Artificial Intelligence (AI) has significantly transformed the landscape of higher education, revolutionizing the way students interact with and learn from educational resources. These innovations have enabled the development and deployment of mobile and intelligent applications, creating new opportunities for personalized and flexible learning experiences with far-reaching implications for the future of higher education (Machado et al., 2024). The widespread adoption of mobile technology, in particular, has facilitated equitable access to educational tools by breaking down barriers related to geography and socioeconomic status. Mobile applications now provide diverse functionalities such as e-books, interactive simulations, educational games, and communication tools, offering convenience and flexibility that especially benefit non-traditional students, including working professionals and those with family responsibilities (Husnita et al., 2023). By enabling students to access a wealth of knowledge anytime and anywhere, mobile learning empowers learners to study independently and at their own pace, thereby supporting more inclusive and adaptable educational pathways.

Furthermore, by offering individualized learning experiences catered to each student's needs, AI-driven applications are revolutionizing higher education. In order to provide personalized learning pathways that maximize the educational experience, artificial intelligence (AI) algorithms examine enormous volumes of data, including behavioral patterns, learning preferences, and student success indicators. These intelligent systems can identify knowledge gaps, recommend targeted resources, and adjust the pace of learning to suit each student's progress (Aditi, 2022). As a result, learners receive a more personalized education that not only enhances comprehension and retention but also fosters a deeper connection with the subject matter. Moreover, AI-driven applications facilitate real-time feedback and assessment, enabling students to promptly address their weaknesses and reinforce their learning.

The integration of mobile and intelligent applications in higher education also fosters a collaborative and connected learning community. Social learning platforms and communication tools embedded within these applications enable students to engage with peers and instructors beyond the confines of the traditional classroom. Collaborative features, such as discussion forums, group projects, and peer review systems, encourage active participation and knowledge sharing, essential for developing critical thinking and problem-solving skills (Salhab & Daher, 2023). Furthermore, the ability to connect with a global network of learners and experts enriches the educational experience, exposing students to diverse perspectives and innovative ideas. This interconnectedness enables students to navigate complex problems and develop solutions that are informed by multiple viewpoints.

The future of learning will be significantly shaped by the thoughtful integration of mobile and intelligent technologies as higher education continues to change. Educators and institutions may build a more personalized, inclusive, and connected learning environment that meets the different needs of students by utilizing the potential of these technologies. Prioritizing innovation, teamwork, and adaptability will be crucial as we go forward to make sure that higher education stays current and can adapt to the shifting demands of society (Quadri et al., 2023). By doing this, we can fully utilize intelligent and mobile technologies, enabling students to thrive in a world that is becoming more sophisticated and linked by the day.

Here are the objectives of this conceptual study.

1. To explore the transformative impact of mobile technology and intelligent applications in higher education, focusing on how these innovations enhance learning experiences and optimize educational outcomes.
2. To examine the integration of AI-driven applications to provide personalized learning paths tailored to individual student needs, thereby improving engagement and performance metrics.
3. To evaluate the role of collaborative and interactive learning environments facilitated by mobile applications, emphasizing their contribution to developing essential 21st-century skills like teamwork and digital literacy.

## METHODOLOGY

This book chapter explores the optimization of higher education learning through mobile and intelligent applications by examining secondary data from academic journals, industry reports, and existing research studies. Relevant literature on educational technology and mobile learning will be reviewed to identify key trends and advancements. Along with case studies and success stories, the prospects and challenges of sustainable financing were discussed.

## EVOLUTION OF MOBILE EDUCATION

### Timeline

The journey of mobile education has been a remarkable one, evolving from simple mobile phones and PDAs to today's sophisticated smartphones, tablets, and laptops. Over the years, mobile learning has revolutionized the way we access and experience education, offering unparalleled flexibility, convenience, and personalization. From basic text messages and SMS quizzes to immersive AR/VR experiences and AI-driven adaptive learning, mobile education has made significant progress, transforming the lives of students, teachers, and learners across the globe. As we continue to embrace new technologies and innovations, mobile learning will undoubtedly stay at the leading edge of education, influencing how people will study in the future and enabling them to realize their greatest potential. (Abid Haleem et.al,2022)

## Early 1990s: Introduction of Mobile Phones

The introduction of mobile phones in the early 1990s revolutionized communication, making it portable and convenient. Although these early devices were primarily used for voice calls and text messages, they laid the groundwork for the future of mobile learning by introducing the concept of technology on the go. Educators and technologists began to envision the potential of these portable devices for delivering educational content and facilitating learning outside traditional classroom settings (Halder et al., 2015).

## Late 1990s: Rise of Personal Digital Assistants (PDAs)

In the late 1990s, Personal Digital Assistants (PDAs) like Palm Pilot and Pocket PC emerged, offering basic computing capabilities such as note-taking, scheduling, and accessing digital content. These devices were the precursors to modern smartphones and tablets, providing early examples of how mobile technology could be used for educational purposes. Educators experimented with using PDAs to distribute course materials, manage assignments, and enhance student organization (Halder et al., 2015).

## Early 2000s: Advent of Mobile Learning (M-Learning)

The early 2000s saw the advent of mobile learning (m-learning) as mobile phones with internet capabilities became more widespread. Educational institutions began to explore the potential of mobile devices for delivering learning content anytime, anywhere. Early m-learning initiatives included

SMS-based quizzes, mobile-friendly websites, and basic educational apps that gave students access to study materials on their phones. (Caudill, 2007)

*Figure 1. Timeline of evaluation of mobile education*

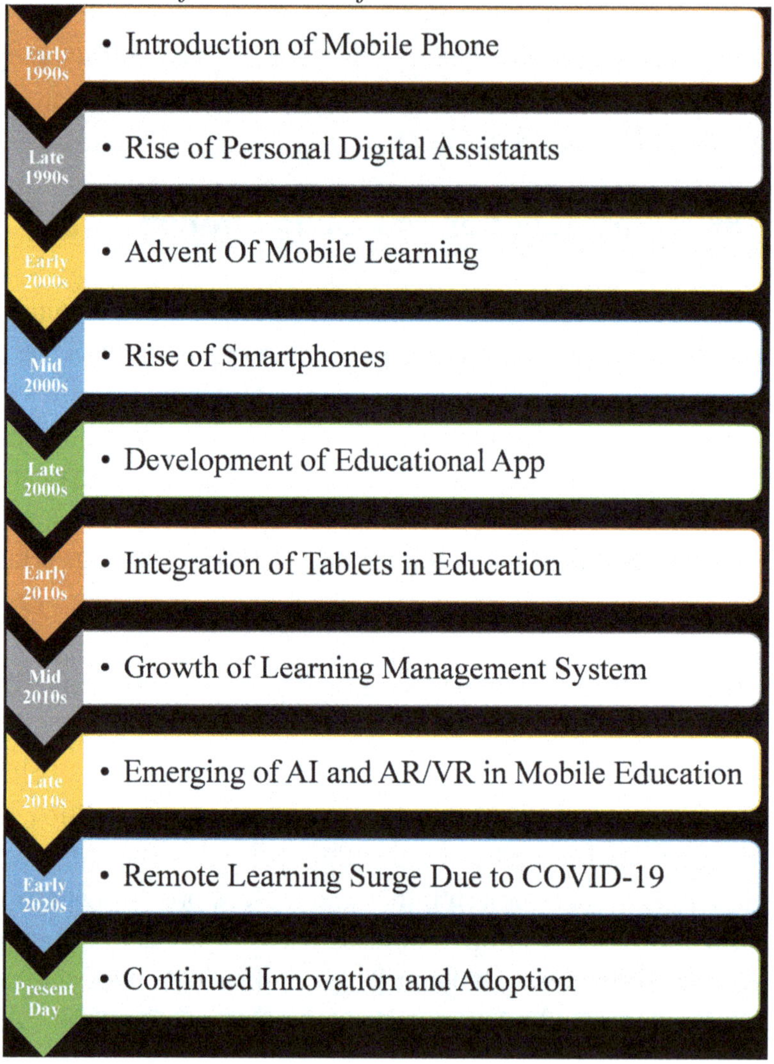

## Mid-2000s: Rise of Smartphones

The mid-2000s marked a significant milestone with the rise of smartphones, most notably the launch of the Apple iPhone in 2007 and the Google Android platform in 2008. These devices featured touchscreens, powerful processors, and access to a wide range of applications, making them ideal for mobile learning. The availability of app stores opened up new possibilities for educational apps that offered interactive and engaging learning experiences (Teodorescu et al., 2023).

## Late 2000s: Development of Educational Apps

In the late 2000s, educational apps began to proliferate, providing students with interactive and engaging ways to learn. Apps like Duolingo for language learning, Khan Academy for various subjects, and Evernote for note-taking became popular tools for students of all ages. These apps leveraged the capabilities of smartphones to deliver rich multimedia content, interactive exercises, and personalized learning experiences. (Hidayati & Diana, 2019)

## Early 2010s: Integration of Tablets in Education

The early 2010s saw the integration of tablets into education, with devices like the iPad offering larger screens and more robust capabilities for educational purposes (Liu, 2022). Schools and universities started adopting tablets for digital textbooks, interactive lessons, and multimedia content. Tablets became valuable tools for classroom instruction, enabling teachers to present content in dynamic ways and allowing students to interact with learning materials through touch and gesture-based interfaces.

## Mid-2010s: Growth of Learning Management Systems (LMS)

In the mid-2010s, Learning Management Systems (LMS) such as Blackboard, Canvas, and Moodle integrated mobile apps, enabling mobile device access to educational content, assignment submission, and discussion participation for students. These systems facilitated seamless communication between students and instructors, streamlined the distribution of course materials, and enabled real-time assessment and feedback, enhancing the overall learning experience. (Bradley, 2021)

## Late 2010s: Emergence of AI and AR/VR in Mobile Education

In the late 2010s, Artificial Intelligence (AI) and Augmented Reality/Virtual Reality (AR/VR) technologies began to be incorporated into mobile educational apps. AI-driven tutors provide individualized instruction by evaluating student data and modifying the material to meet each student's needs (Dembe & Extension, 2024). AR/VR offered immersive simulations and interactive content, enabling students to participate in interactive learning activities and visualize difficult topics in virtual settings.

## Early 2020s: Remote Learning Surge Due to COVID-19

The COVID-19 pandemic in 2020 led to a significant surge in remote learning as schools and universities worldwide adopted online learning platforms to continue education amidst lockdowns. Mobile education became crucial during this period, with video conferencing programs like Microsoft Teams and Zoom becoming as indispensable resources for online learning environments (Dhawan, 2020). The epidemic emphasized the value of adaptability and accessibility in education while also hastening the introduction of mobile learning tools.

## Present Day: Continued Innovation and Adoption

Today, mobile education continues to evolve with ongoing innovations. Educational apps leverage AI for personalized learning, AR/VR for immersive experiences, and gamification to engage students. The focus is on creating inclusive, flexible, and engaging learning environments accessible to all students. Mobile education has become an integral part of the educational landscape, giving pupils the instruments and materials need for success in a world that is changing quickly (Rizvi, 2024).

## ENHANCEMENTS BROUGHT BY MOBILE APPLICATIONS IN HIGHER EDUCATION

### Convenience and Adaptability

Convenience and Adaptability in higher education have been revolutionized by mobile applications, transforming the learning experience into a more accessible, flexible, and personalized journey. These innovative apps have shattered the constraints of traditional classroom settings and rigid schedules, allowing students to access educational resources anywhere, anytime. This adaptability caters to diverse learning styles and paces, tailoring the experience to individual needs and preferences. Students can now review lecture notes, complete assignments, or engage in discussions while on the move, making education more convenient and inclusive (Hinze et al., 2022).

### Personalized Learning

Personalized Learning through mobile applications takes education to the next level by tailoring the experience to each student's unique needs and preferences. By harnessing the power of data analytics and AI, these apps create customized learning paths, offering resources and exercises that match each student's pace and style (Jian, 2023). This means students can focus on areas where they need improvement, skipping what they already know, and accelerating their learning journey. For instance, a language learning app might provide additional practice with challenging grammar rules for some students while offering advanced vocabulary for others. This personalized approach keeps students engaged, as the content is relevant and appropriately challenging, leading to enhanced understanding and retention.

## Collaborative Learning

Collaborative Learning through mobile applications enables learners collaborate with one another seamlessly, regardless of their location. These apps provide features like group chats, shared documents, and discussion forums, facilitating effortless communication and teamwork. Students can collaborate on projects, share ideas, and give feedback in real-time, promoting a sense of community and enhancing learning. For example, group assignments become more manageable, and brainstorming sessions can happen anytime, anywhere. Collaborative learning also helps develop essential skills like communication, cooperation, and problem-solving, making learning more interactive and engaging (Papadopoulos et al., 2018).

## Engagement and Motivation

Engagement and Motivation through mobile applications elevate the learning experience by making it interactive, enjoyable, and rewarding. Features like quizzes, games, and simulations keep students interested and actively participating, while rewards like points or badges motivate them to complete tasks and improve. Interactive content, such as videos and animations, makes learning fun and easier to understand. By providing immediate feedback, mobile applications help students stay on track and feel a sense of achievement, leading to better retention and comprehension of the material (Tejada-Simon, 2024)

## Real-time Feedback and Assessment

Real-time Feedback and Assessment through mobile applications significantly enhance the learning process by providing immediate insights into students' performance. These apps allow students to complete quizzes and assignments and instantly see their results, helping them identify areas that need improvement right away. For example, if a student gets a math problem wrong, the app can explain the mistake and offer additional practice problems. This immediate feedback keeps students engaged and helps them learn more effectively. Teachers also benefit, as they can monitor progress in real-time and quickly address any issues, ensuring students stay on track and achieve their learning goals (Horvers et al., 2024).

### Resource Efficiency

Resource Efficiency through mobile applications reduces the need for physical materials like textbooks and paper, making education more cost-effective and environmentally friendly. Digital resources, such as e-books and online articles, save money and trees, while mobile apps streamline administrative tasks, allowing educators to devote more time to instructional activities (Chilaka, 2024). Students can access a wealth of information without the need for extensive physical libraries, making learning more resource-friendly and economical.

### Remote Learning Capabilities

Remote Learning Capabilities of mobile applications have revolutionized education, especially during times of crisis like the COVID-19 pandemic. These apps enable students to attend classes, access materials, and complete assignments from anywhere, ensuring learning continues uninterrupted. Video conferencing tools like Zoom and Microsoft Teams allow for virtual classrooms where students and teachers can interact in real-time (Abed & Alsalih, 2023). Recorded lectures and online resources make it easy for students to review content at their convenience. Discussion forums and chat features keep students linked and involved with their classmates, but applications for distant learning support various learning styles, catering to individual needs and providing flexibility to balance education with other commitments.

## REVOLUTIONIZING HIGHER EDUCATION: THE POWER OF LMS AND EDUCATIONAL APPS

In the swiftly changing terrain of higher education, Learning Management Systems (LMS) such as Blackboard, Canvas, and Moodle have become indispensable tools, revolutionizing the way courses are structured and presented. These platforms offer a unified digital space where course resources, assignments, grades, and feedback are readily available, enabling effortless communication between students and instructors. In conjunction with LMS, subject-specific applications and productivity tools are further augmenting the educational experience by providing customized features that cater to diverse academic fields and promote effective learning strategies (Ymchenko, 2022). These technological innovations not only enhance the accessibility and adaptability of learning but also create a more immersive and dynamic educational setting. By combining LMS and specialized educational apps, higher education institutions can more effectively cater to varied learning preferences, simplify administrative tasks for instructors, and ultimately improve student performance.

## Learning Management Systems (LMS)

Learning Management Systems (LMS) have become the cornerstone of modern higher education, serving as the foundation for digital learning environments. LMS platforms, such as Blackboard, Canvas, and Moodle, offer comprehensive solutions that streamline the management of educational courses and facilitate seamless communication between students and educators. (Oluwaseyi & Cena, 2024). At their core, LMS platforms offer a central location for homework assignments, course materials, grades, and feedback can be accessed and managed efficiently, making it easier for students and educators to stay organized and connected.

LMS offers unusual accessibility and flexibility, that is one of its key benefits. Students can access their course content anytime and anywhere, which is particularly beneficial for those who juggle studies with work or personal commitments. This on-demand access to learning resources supports diverse learning styles It makes learning more flexible and inclusive by letting students go at their own speed (Furqon et al., 2023). For instance, recorded lectures and downloadable materials enable students to review content as many times as necessary to grasp complex concepts fully, while also catering to different learning preferences

LMS platforms also enhance the organization of courses and the administration of educational activities, making it easier for educators to manage their workload. Educators can easily upload syllabi, schedule assignments, create quizzes, and track student progress all in one place, saving time and effort. Automated grading features save instructors' significant time and effort, providing pupils with immediate feedback so they may determine what needs to be improved. This real-time assessment capability is crucial for maintaining a continuous and responsive learning environment, ensuring that students receive timely support and guidance (Inyutina & Kartashova, 2023).

Furthermore, LMS fosters collaboration and communication within the academic community, extending beyond the physical classroom. Features such as discussion forums, group projects, and messaging systems facilitate interaction between students and teachers as well as among peers, encouraging active participation and engagement. These tools support collaborative learning, cultivating a feeling of belonging and teamwork that enhances the overall learning experience (Mealings et al., 2024). Integrated tools like Google Workspace and Microsoft Office 365 further support collaborative learning, allowing students to co-author documents and presentations seamlessly.

The integration of multimedia content is another notable feature of LMS platforms, making learning more dynamic and engaging. Videos, interactive simulations, and other multimedia resources can be embedded within the course content, catering to various learning preferences and helping students understand and retain information more effectively. For example, science courses can incorporate virtual lab experiments, while history classes might include interactive timelines and maps, providing a more immersive learning experience (Aggarwal et al., 2024).

Security and privacy are critical aspects of LMS platforms, ensuring that sensitive information such as grades and personal data are protected through robust security measures (Jose, 2024). Institutions can control access to content guaranteeing that data may only be viewed or altered by authorized individuals, maintaining academic integrity and protecting the privacy of both students and faculty.

In addition to these functionalities, many LMS platforms are continually evolving to incorporate sophisticated technology like analytics and artificial intelligence (AI). By seeing patterns and trends, these improvements can offer deeper insights into student performance and give useful information. teaching strategies and improving educational outcomes. For example, AI-powered tools can predict which students might need additional support based on their interaction with the platform and their performance metrics, enabling early intervention and targeted support (Sougleridi et al, 2024).

The adaptability of LMS platforms is another significant benefit, allowing them to cutting edge technology like analytics and artificial intelligence (AI). By detecting patterns and trends in student performance, these improvements can offer deeper insights that can guide institutions, programs, and courses. Whether it's a small community college or a large university, an LMS can be tailored to support various educational models, including blended learning, flipped classrooms, and fully online programs. This versatility ensures that LMS platforms can evolve alongside changing educational trends and demands, providing a future-proof solution for institutions (Nasir et al., 2023).

## Subject-Specific Apps

Subject-specific apps have become indispensable in higher education, offering targeted tools and resources tailored to particular academic disciplines. These applications address the unique needs and challenges of various fields of study, providing specialized support that surpasses general educational apps (Oluwaseyi & Cena, 2024). For instance, apps developed for subjects like mathematics, science, or languages offer focused functionalities that enhance learning and mastery in those areas.

In mathematics, apps such as Photomath and Wolfram Alpha provide students with tools to solve complex problems, visualize equations, and understand mathematical concepts in a more interactive and immersive way. These apps allow students to take pictures of handwritten equations or type them in to receive step-by-step solutions and explanations, thereby aiding in homework completion and concept reinforcement (Silvestre & Jacinto, 2021). Similarly, apps like Khan Academy offer comprehensive lessons on a broad range of mathematical subjects, including sophisticated calculus and simple arithmetic interactive exercises and instructional videos catering to various skill levels and learning styles.

For science disciplines, subject-specific apps offer simulations, virtual labs, and interactive models that enable students to explore scientific phenomena and conduct experiments in a virtual environment. Apps like Labster provide immersive, computer-based simulations that replicate real laboratory experiences, allowing students to perform experiments safely and effectively. These virtual labs can be particularly useful for students in fields such as chemistry and biology, where physical lab access may be limited or restricted. Interactive apps also allow students to visualize complex processes, such as cellular functions or chemical reactions, making abstract concepts more tangible, accessible, and easier to grasp (Polat & Gödek, 2021).

Language learning apps are another example of subject-specific applications that cater to the needs of students studying foreign languages. Apps like Duolingo and Babbel use gamified approaches to teach vocabulary, grammar, and pronunciation through interactive exercises and engaging activities. These apps often incorporate speech recognition technology to help students practice and improve their pronunciation and listening skills, providing personalized feedback and assessment (Kessler et al., 2023). By offering personalized learning paths and adaptive content, language learning apps provide a flexible and effective way for students to learn a fresh language at their own pace, anytime and anywhere.

Subject-specific apps also benefit educators by providing tools for creating customized learning experiences. For example, an educator teaching a course in physics might use a simulation app to demonstrate complex concepts like wave interference or quantum mechanics, allowing students to visualize these phenomena and engage with the material in a more meaningful and interactive way (Meccawy, 2022). These apps can be integrated into the curriculum to enhance conventional teaching strategies, providing lively and participatory means to reinforce key concepts and promote deeper understanding.

Moreover, subject-specific apps often include features such as progress tracking, quizzes, and assessments that help students monitor their learning and identify areas for improvement. For example, apps like Quizlet allow students to create and use digital flashcards for studying and testing their knowledge on specific topics, providing a personalized and self-directed learning experience (Swailes, 2016). This ability to assess and track progress in real-time supports a more personalized and adaptive learning approach, empowering pupils to assume responsibility for their education journey.

The integration of subject-specific apps into higher education also aligns with the growing emphasis on personalized and adaptive learning. By providing resources and tools tailored to individual subjects, these apps support diverse learning styles and preferences, making education more accessible, effective, and inclusive for a wide range of students. They also enable students to engage with content in a more interactive, practical, and relevant manner, bridging the disconnect between theoretical comprehension and practical application, and setting up students for success in their chosen fields (Joshi et al., 2023).

## Study and Productivity Apps

Study and Productivity Apps have become indispensable tools in higher education, revolutionizing the way students manage their studies and optimize their productivity. These apps are designed to help students streamline their coursework, organize their study routines, and master their time management skills (Croucher, 2019). By leveraging these apps, students can significantly enhance their ability to stay on top of their academic responsibilities and achieve their goals.

Note-taking apps, such as Evernote and OneNote, empower students to capture and organize lecture notes, research, and personal reminders in a single, centralized location. These apps often feature advanced organizational tools, including tagging, searching, and multimedia integration, making it easier for students to retrieve and review information efficiently (Cojean & Grand, 2024). This enables students to focus on understanding and applying the material rather than wasting time searching for scattered notes

Time management apps, such as Todoist and Google Calendar, assist students in scheduling and prioritizing tasks, ensuring they stay on track and meet deadlines (Ariarta et al., 2024). By setting reminders and creating task lists, students can maintain a structured schedule, balance academic responsibilities with other commitments, and avoid last-minute cramming or missed deadlines. This leads to a more organized, less stressful academic experience, allowing students to perform at their best.

Productivity apps also include tools designed to enhance focus and concentration. Apps like Focus@Will and Forest help students minimize distractions and stay on task. Focus@Will offers music specifically designed to enhance concentration, while Forest encourages students to stay focused by growing a virtual tree that flourishes when they avoid using their phones (Agasi, 2024). These apps support students in developing effective study habits, improving their overall productivity, and achieving a better work-life balance.

Study aid apps, such as Quizlet and Anki, provide interactive ways to review and memorize information, making the study process more engaging and tailored to individual learning styles. Quizlet allows students to create and use digital flashcards, while Anki uses spaced repetition algorithms to help students retain information more effectively. These apps offer a variety of study modes, including practice quizzes and games, enabling students to reinforce their understanding and retain information better (Indra et al., 2023).

The integration of study and productivity apps into students' academic lives also supports collaborative learning. Tools like Google Docs and Microsoft Office 365 enable real-time group project and task collaboration that enables students to cooperate without difficulty despite being physically apart (Muhammad Nazeef et al., 2024). The cooperation, communication, and problem-solving abilities that these collaborative elements promote are critical in both academic and professional contexts.

By embracing study and productivity apps, students can optimize their academic experience, achieve greater success, and develop essential skills for their future careers. These apps provide a personalized, flexible, and effective way to manage studies, stay organized, and maintain productivity, ultimately leading to a more fulfilling and accomplished academic journey.

## CHALLENGES AND CONSTRAINTS IN MOBILE LEARNING

The integration of smart and mobile applications in higher education has undoubtedly yielded substantial benefits and improvements, revolutionizing how pupils engage with and learn from instructional materials. Nonetheless, it's equally critical to recognize and comprehend the array of challenges and limitations that accompany the adoption of these technologies in academic settings (Qasim, 2024). By doing so, educators and administrators can effectively harness the potential of smart and mobile applications to enhance the learning experience, while concurrently mitigating their negative consequences and minimizing potential drawbacks. This informed approach ensures that the benefits of technology integration are maximized, and its limitations are addressed, ultimately creating a more effective and supportive educational environment.

### Digital Divide and Accessibility

One of the most significant hurdles in Incorporation of smart and mobile applications in higher education is the pervasive issue of the digital divide, which represents the stark disparity between individuals who possess seamless access to cutting-edge information and communication technology, and those who are left behind without such access. A substantial number of students lack reliable internet connectivity or the requisite devices, including smartphones, tablets, and laptops, necessary to fully leverage the benefits of smart and mobile applications. (Qasim, 2024). This, pronounced disparity has far-reaching consequences, potentially exacerbating existing educational inequalities by placing disadvantaged students at a significant disadvantage in comparison to their classmates' enjoy unfettered access to technology. It is imperative to address this issue by ensuring equitable access to digital resources, thereby bridging the gap and preventing the widening of existing educational disparities, ultimately promoting a more inclusive and equitable learning environment

## Technical Issues and Reliability

The efficacy of smart and mobile applications in higher education can be significantly impeded by a multitude of technical issues, including but not limited to, software glitches, connectivity disruptions, and device compatibility problems. Students and educators alike may encounter a plethora of difficulties, such as application crashes, sluggish loading times, and inconsistent performance, which can severely disrupt the learning process, lead to heightened frustration levels, and ultimately undermine the overall educational experience. Furthermore, the rapid pace of technological innovation necessitates frequent updates to ensure the continued functionality and security of apps, which can pose a substantial logistical challenge for both users and developers, requiring significant time, resources, and expertise to maintain and troubleshoot, thereby adding an extra degree of intricacy to the integration of smart and mobile applications in academic settings (Mesuwini & Mokoena, 2024).

## Security and Privacy Concerns

The incorporation of smart and mobile applications in higher education inherently entails the collection, storage, the examination of enormous volumes of delicate personal data, encompassing academic records, personal identifiable information, and usage patterns. As such, ensuring the unwavering security and privacy of this data is a paramount concern that necessitates meticulous attention from educational institutions and app developers alike. To safeguard against potential data breaches, unauthorized access, and cyber threats, robust and multi-layered security measures must be implemented, including but not limited to, encryption, secure authentication protocols, and regular security audits. Furthermore, the use and sharing of student data raise critical concerns of openness, informed consent, and the possibility of exploitation, underscoring the need for strict adherence to stringent data protection regulations, ethical guidelines, and industry best practices. By giving data security and privacy first priority, educational institutions can foster trust, maintain academic integrity, and ensure a safe and secure learning environment for all students. (Kumi-Yebaoh et.al,2023)

## Pedagogical Limitations

While smart and mobile applications offer groundbreaking opportunities to captivate and engage students, their innovative nature can sometimes pose challenges for educators who are steeped in traditional pedagogical approaches. Some instructors may struggle to seamlessly integrate these technologies into their established teaching methods, or might need more assistance and training to harness their full potential. Moreover, there exists a significant risk that an overemphasis on technology could inadvertently eclipse fundamental educational principles, leading to an undue reliance on apps and digital tools at the expense of critical thinking, deep learning, and nuanced understanding. To mitigate this risk, it is essential to strike a delicate balance between technological tools and effective pedagogical strategies, ensuring a holistic and comprehensive educational experience that leverages the strengths of both traditional and innovative approaches (Modlinger, 2024). By doing this, teachers may provide a stimulating and dynamic learning environment that encourages creativity, intellectual curiosity, and academic success.

## Distraction and Over-Reliance on Technology

Mobile devices and applications, while offering immense educational potential, can also serve as a significant source of distraction for students, posing a formidable challenge to their focus and productivity. The very same devices utilized for academic purposes can simultaneously provide effortless access to a vast array of non-educational content, including social media, games, and other digital diversions. This dual nature of mobile technology can have a profoundly negative impact on students' concentration, attention span, and overall academic performance. Furthermore, there is a pressing concern that excessive reliance on technology may diminish students' capacity for independent learning, critical thinking, and problem-solving skills. To effectively mitigate these issues, it is crucial to promote responsible technology use, foster digital literacy, and encourage students to develop healthy relationships with their devices, thereby ensuring that technology serves as a tool to augment, rather than hinder, their educational journey (Lavidas et al., 2020).

## Cost and Resource Allocation

The integration of smart and mobile applications in higher education necessitates a substantial financial commitment, presenting a significant challenge for institutions with limited budgets. To successfully implement these technologies, colleges and universities must allocate considerable resources for the procurement of devices, development or subscription to applications, and establishment of the requisite infrastructure, including servers, networks, and security systems. Furthermore, institutions must also factor in ongoing expenses related to staff training, software updates, technical support, and maintenance, which can add up quickly. For some institutions, particularly those with constrained financial resources, these monetary constraints can pose a formidable barrier to adopting new technologies, hindering their ability to provide students with the benefits of mobile learning and potentially exacerbating existing disparities in access to quality education (Yang & Zhong, 2023). As such, careful budgeting, strategic planning, and exploration of cost-effective solutions are essential to ensure that financial constraints do not impede the adoption of innovative technologies.

## Content Quality and Standardization

The quality and effectiveness of smart and mobile applications can vary significantly, with a vast disparity in the educational value and academic rigor they offer. Unfortunately, not all educational apps are created with the same level of expertise, attention to detail, and commitment to excellence, resulting in some apps providing subpar content that fails to meet stringent academic standards. Ensuring that the content provided by these applications is accurate, relevant, engaging, and of exceptionally high quality is crucial to maintaining the integrity of the educational experience (Husnita et al., 2023). Moreover, there is a pressing need for standardization and consistency in the development and evaluation of educational apps to ensure that they align seamlessly with curriculum requirements, learning objectives, and pedagogical best practices. Educators and institutions must therefore exercise diligence and discernment in carefully evaluating and selecting applications that not only improve the learning environment but also support and advance their academic goals, foster critical thinking, and promote intellectual curiosity. By doing so, they able to fully utilize mobile learning to enrich student outcomes and academic achievement.

## Teacher Training and Professional Development

The successful integration of smart and mobile applications in educational settings necessitates that educators possess a high level of proficiency in utilizing these technologies to enhance teaching and learning. However, a significant number of teachers may lack the requisite training, support, and expertise to effectively incorporate these digital tools into their pedagogical practices, potentially hindering the realization of their full potential. To address this challenge, it is essential to provide educators with comprehensive, sustained, and high-quality professional development programs that equip them with the necessary skills, knowledge, and confidence to navigate the complexities of digital learning tools. Additionally, ongoing support, mentoring, and coaching are crucial to help educators stay abreast of emerging technologies, troubleshoot technical issues, and maximize the benefits of mobile learning. By investing in educator capacity-building and providing a supportive environment, institutions can empower teachers to harness the potential of smart and mobile applications, foster innovative pedagogies, and improve student outcomes (Dahri et al., 2024).

## Impact on Social Interaction

While mobile applications offer unparalleled opportunities to facilitate communication and collaboration, there exists a legitimate concern that excessive reliance on technology may inadvertently erode the quality and frequency of face-to-face social interactions, which are essential for building strong, meaningful relationships and a sense of community. The educational experience extends far beyond academic knowledge, because it's critical to acknowledge the significance of fostering interpersonal skills, empathy, and social competence in students. Therefore, it is imperative to strike a delicate balance between digital and in-person interactions, ensuring that students have ample opportunities to engage in rich, nuanced, and face-to-face exchanges with their peers and instructors (Ghoulam et al., 2024). By doing so, educators can help students develop the essential social skills, emotional intelligence, and deep human connections that are vital for success in both academic and professional spheres, while also fostering a welcoming and inclusive learning atmosphere that honors each individual's distinctive contributions.

## CASE STUDIES AND SUCCESS STORIES

Mobile learning has revolutionized higher education, yielding remarkable success stories across diverse institutions, and demonstrating its capacity to improve learning results, motivation, and student involvement. Notable examples showcase mobile learning's transformative impact on academic achievement and student satisfaction, while institutions worldwide are leveraging mobile technologies to create interactive, personalized, and inclusive learning environments. These success stories offer valuable insights and best practices for educators seeking to enhance student success. By embracing mobile learning, institutions can promote academic innovation and excellence, ultimately transforming the educational landscape. The impact of mobile learning is far-reaching, with the potential to increase access, equity, and quality in higher education. As mobile learning keeps changing, it's imperative to share and build upon these success stories to maximize their impact.

### Stanford University's iPad Initiative

Stanford University launched a ground-breaking program that provided medical students with iPads loaded with custom apps and digital textbooks. The program's audacious aim was to transform the conventional educational model by substituting dynamic, interactive digital content for static textbooks. This innovative method made it easy to obtain the most recent data, interactive teaching aids, and multimedia materials, which promoted a more immersive and interesting learning environment. Due to the learning materials' mobility and accessibility, students were able to study both on and off campus at their own speed and from any location. This resulted in an astounding increase in student engagement and academic success. This project not only improved the quality of instruction overall but also demonstrated how digital technology can change medical education and make it more student-centered, efficient, and accessible. Stanford University created a new standard for creative pedagogy by utilizing the potential of mobile learning, opening doors for other educational institutions to follow.(www.stanford.edu)

## Duke University's Mobile Device Program

Duke University pioneered a groundbreaking mobile device program, where incoming students were presented with iPods preloaded with a comprehensive array of orientation materials, course content, and innovative educational apps, with the ambitious objective of seamlessly integrating mobile technology into the academic curriculum and significantly enhancing the learning environment. This visionary initiative was met with overwhelmingly positive feedback from students, who reported a notable increase in engagement, motivation, and ease of access to a vast array of educational resources, thereby facilitating a more immersive and interactive learning experience. The resounding success of this program served as a catalyst for the widespread adoption of mobile technologies across various courses and departments, further solidifying Duke University's reputation as a trailblazer in the realm of innovative pedagogy and academic excellence. By harnessing the power of mobile learning, Duke University created a dynamic and inclusive learning ecosystem that continues to inspire and empower students to reach their full potential. (https://duke.edu)

## University of Central Florida's Mobile Learning Initiatives

The University of Central Florida (UCF) has been a pioneer and leader in the innovative integration of mobile learning into its curriculum, demonstrating a steadfast commitment to harnessing the potential of mobile technologies to enhance student learning outcomes. UCF has developed a range of mobile-friendly courses, leveraging cutting-edge apps like Canvas for comprehensive course management, which enables students to effortlessly access course materials, submit assignments, and engage in seamless communication with instructors via their mobile devices. This forward-thinking approach has yielded a significant increase in student participation, engagement, and motivation, while also making learning more accessible, flexible, and convenient, particularly for non-traditional students who balance multiple responsibilities, including work, family, and education. By embracing mobile learning, UCF has created a more inclusive, student-centered, and dynamic learning environment that caters to the diverse needs of its student population, setting a new standard for academic excellence and innovation in higher education.(www.ucf.edu)

## Open University UK's Use of Mobile Technologies

The Open University in the UK has achieved remarkable success in seamlessly integrating mobile learning into its distance education programs, revolutionizing the way students access and engage with course materials. By leveraging cutting-edge mobile apps and platforms, the university has provided students with unparalleled flexibility and autonomy, enabling them to access a vast array of course materials, participate in interactive activities, and engage in online discussions with peers and instructors at their convenience, regardless of their geographical location. This innovative approach has yielded a significant improvement in student retention and satisfaction rates, as learners can now engage with their studies in a more personalized, self-paced, and flexible manner, fostering a deeper sense of connection and community. By harnessing the power of mobile learning, the Open University has set a new benchmark for distance education, demonstrating its commitment to providing inclusive, accessible, and high-quality learning experiences that cater to the diverse needs of its students. (www.open.ac.uk)

## Arizona State University's Smart Campus Initiative

Arizona State University (ASU) pioneered a groundbreaking Smart Campus initiative, harnessing the potential of mobile apps to revolutionize student life and learning, and create a more connected, inclusive, and supportive campus community. ASU developed a bespoke mobile app that served as a one-stop-shop for students, providing them with real-time information about campus events, academic deadlines, transportation schedules, and other vital resources. Furthermore, the app seamlessly integrated with the university's learning management system, empowering students to access course materials, grades, and other academic tools on-the-go, thereby streamlining their learning experience. This innovative initiative yielded a significant improvement in campus connectivity, student engagement, and academic performance, as students were able to navigate campus life with greater ease, convenience, and confidence. By leveraging mobile technology, ASU has set a new standard for student-centric innovation, demonstrating its commitment to fostering a dynamic, responsive, Creating a learning atmosphere that is focused on the needs of the students and equips them for success in the digital era. (www.asu.edu)

## Penn State University's Mobile Technology Integration

Penn State University has been at the forefront of embracing mobile technology, pioneering the integration of innovative educational apps and platforms into its teaching methodologies, and revolutionizing the way students learn and engage with course materials. A notable example of this is the university's adoption of Poll Everywhere, a cutting-edge tool that enables instructors to conduct real-time surveys and quizzes during lectures, thereby increasing student participation, fostering a sense of community, and providing immediate feedback that informs instruction. This strategic integration of mobile technology has given rise to a more dynamic, interactive, and immersive learning environment, characterized by heightened student engagement, motivation, and success rates. By harnessing the power of mobile technology, Penn State University has created a student-centered ecosystem that supports academic excellence, creativity, and critical thinking, ultimately preparing students to thrive in an increasingly complex and interconnected world. (www.psu.edu)

## FUTURE TRENDS AND OUTLOOK

Mobile education will transform significantly in the coming years due to technological advancements and evolving educational needs. Key trends include the use of AI and ML for personalized learning, enhanced AR and VR for immersive experiences, and microlearning for quick, focused sessions. Improved collaboration tools, gamified learning, and blockchain for credentialing and security will also play crucial roles. The rollout of 5G, integration with smart wearables, and a focus on digital well-being will further shape the future of mobile education (Dembe & Extension, 2024).

*Table 1. Emerging trends in mobile education*

| Emerging Trends | Description |
|---|---|
| AI and Machine Learning | AI and ML will revolutionize learning, analyzing patterns to adapt content to individual needs. Intelligent tutoring systems and chatbots will provide instant assistance and tailored study plans, offering personalized learning experiences that cater to unique learning styles and goals. |
| AR and VR Integration | AR and VR technologies will revolutionize immersive learning, rendering complex subjects more comprehensible and captivating. AR superimposes digital information on physical objects, whereas VR generates virtual environments for simulations and hands-on exercises, fostering a more engaging and interactive learning experience. |
| Microlearning and Bite-sized Content | Microlearning will provide educational content in concise, bite-sized units, perfectly suited for mobile, on-the-go learning. Short, engaging videos, quizzes, and interactive modules will facilitate swift, concentrated learning sessions, accommodating busy schedules and diverse learning styles. |
| Improved Collaboration Tools | Advanced mobile apps will provide enhanced collaboration tools, facilitating seamless group projects, instant communication, and peer-to-peer learning through dynamic, interactive platforms, fostering a more connected and productive learning environment. |
| Gamification of Learning | Gamified learning apps will leverage game design elements to boost engagement and motivation, integrating points, badges, leaderboards, and challenges to transform learning into a more immersive, interactive, and enjoyable experience, making academic success a rewarding and fun pursuit. |

continued on following page

*Table 1. Continued*

| Emerging Trends | Description |
|---|---|
| Blockchain for Credentialing and Security | Blockchain technology will guarantee the integrity and authenticity of digital credentials, safeguarding academic records from tampering and falsification, and streamlining the verification process for employers and institutions, thereby ensuring the validity and trustworthiness of academic achievements. |
| Expanded Offline Capabilities | Mobile learning apps will provide enhanced offline capabilities, enabling students to download educational content and access it without relying on internet connectivity, thereby guaranteeing seamless and uninterrupted learning experiences, even in areas with limited or no internet access. |
| 5G Technology | The deployment of 5G networks will revolutionize mobile learning by delivering significantly faster internet speeds and reduced latency, facilitating the seamless streaming of high-definition video content and enabling more immersive, interactive, and real-time learning experiences. |
| Integration with Smart Wearables | Smart wearables, such as smartwatches and fitness trackers, will be seamlessly integrated into mobile learning platforms, offering innovative methods to interact with educational content and monitor learning progress, thereby enhancing the overall learning experience. |
| Focus on Digital Well-being | Mobile learning platforms will progressively integrate features that foster digital well-being, including tools to monitor and manage screen time, as well as reminders to take regular breaks, thereby ensuring a healthy and balanced approach to technology utilization. |

## AI Driven Tutoring Systems

AI-powered educational platforms could transform the field of education through providing individually tailored learning opportunities. These systems use sophisticated algorithms to evaluate students' learning habits, abilities, and areas of growth, allowing them to provide personalized content and activities. This flexible method enables students to progress at their own speed, making sure they understand each concept thoroughly before advancing. Personalization in education includes immediate feedback from

AI tutors that can correct errors, clarify difficult concepts, and offer extra materials to enhance the learning experience.

Another important benefit of AI-powered tutoring systems is their level of accessibility and availability. AI tutors provide constant support to students at any time and place, unlike traditional tutoring that is restricted by time and location. This level of adaptability is especially advantageous for students with different schedules or those in need of extra assistance beyond typical class times. Having the option to reach educational materials anytime allows students to be more independent and motivated in controlling their learning.

## Integration of Blockchain Technology in Educational Credentials

Incorporating blockchain technology into educational credentials provides a safe and clear method for handling and confirming academic accomplishments. Blockchain's unalterable and secure decentralized nature guarantees that educational credentials, once logged, cannot be modified or counterfeited. This is especially useful in addressing problems such as diploma fraud and verifying the legitimacy of academic records. Students can possess a permanent, convenient digital documentation of their accomplishments, which can be shared with potential employers, schools, or other organizations without requiring outside verification. This not only makes the verification process more efficient but also boosts confidence in the credentials provided.

Furthermore, blockchain technology enables the development of a more interconnected and global educational ecosystem. Using blockchain technology, students have the ability to gather micro-credentials, badges, and certificates from diverse institutions around the globe, all stored on one secure ledger. This can support learning throughout life, enabling individuals to develop a varied range of skills and qualifications that are acknowledged internationally. Incorporating blockchain technology in educational institutions can lessen administrative tasks, enhance data protection, and facilitate the creation of new learning approaches that address the evolving demands of the workforce.

## Future of Mobile Learning With 5G Technology

The future of mobile learning with 5G technology is expected to greatly improve the learning experience through quicker, more dependable, and very interactive educational settings. 5G's enhanced bandwidth and minimal delay enable mobile educational apps to provide live video streaming, top-notch virtual and augmented reality experiences, and smooth communication between students and teachers globally. These progressions enable learners to participate in immersive simulations, virtual labs, and live interactive sessions that were once restricted by technological limitations. As a consequence, 5G will enhance learning experiences that are more interactive and efficient, especially in areas that require hands-on learning and visual representation.

Moreover, the implementation of 5G technology will increase the availability of mobile learning by accommodating various devices and providing connectivity in previously unserviceable areas with high-speed internet. This can aid in closing the gap in digital access, providing quality education to areas with limited resources and enabling students in remote locations to engage in worldwide educational opportunities. Moreover, the capacity to manage numerous connected devices at once enables educational organizations to implement extensive mobile learning systems, ultimately expanding their reach to a greater number of students through personalized content and adaptive learning tools. With the ongoing rollout of 5G networks, the incorporation of this technology into mobile learning will be vital in influencing the future of education.

# DISCUSSION

AI-enabled mobile education is driving a significant transformation in the educational industry by reshaping how knowledge is accessed, delivered, and experienced. Unlike traditional classroom-based methods, students today demonstrate a strong preference for learning through mobile applications, which provide interactive, flexible, and learner-centered environments. This shift highlights the growing acceptance of digital tools as legitimate and effective platforms for education. Moreover, the integration of advanced technologies such as Augmented Reality (AR) and Virtual Reality (VR) into mobile learning ecosystems further enhances the educational experience by allowing students to engage with concepts in simulated, immersive environments. Such innovations not only promote deeper conceptual understanding but also bridge the gap between theoretical knowledge and practical application. Importantly, this transformation fosters inclusivity by enabling learners from diverse backgrounds, including non-traditional students, to access high-quality education anytime and anywhere. However, while the potential of AI-driven mobile education is immense, challenges such as digital literacy gaps, unequal access to devices and internet connectivity, and the need for pedagogical adaptation remain critical considerations. Addressing these issues will be essential for ensuring that the benefits of AI, AR, and VR in mobile education are realized across the wider student community.

## CONCLUSION AND DIRECTIONS FOR FUTURE RESEARCH

The integration of smart and mobile applications in higher education has dramatically transformed the learning landscape, providing enhanced access, personalization, and engagement opportunities. However, it is crucial to recognize and address the accompanying challenges and limitations to maximize the potential benefits and create an effective educational environment. Future studies should focus on a number of important areas to better understand and improve the integration of mobile technology in higher education. Firstly, addressing the digital divide remains a critical challenge, necessitating innovative solutions to ensure equitable access to mobile learning technologies for all students, regardless of their socio-economic background or geographic location. This includes investigating affordable and sustainable methods to provide devices and internet access to underserved populations. Secondly, the development of personalized learning experiences through AI and machine learning should be a priority, with future studies examining the effectiveness of AI-driven applications in tailoring educational content to individual learning styles and needs, optimizing student engagement and performance. This includes evaluating the impact of intelligent tutoring systems and adaptive learning platforms on student outcomes.

Thirdly, the potential of augmented reality (AR) and virtual reality (VR) in creating immersive learning experiences warrants further exploration, with research assessing their effectiveness in enhancing understanding and retention of complex subjects, as well as their ability to engage students in hands-on, experiential learning. Additionally, the rise of microlearning and bite-sized content should be investigated to determine its efficacy in accommodating diverse learning preferences and busy schedules, with studies evaluating the impact of short, focused learning sessions on knowledge retention and academic success. The role of gamification in education also presents an exciting area for research, with future studies exploring how game design elements can be integrated into educational apps to boost motivation and engagement, and assessing their long-term impact on learning outcomes.

Additionally, incorporating blockchain technology for credentialing and security in education should be explored, with research examining how blockchain can ensure the authenticity and integrity of academic records and streamline the verification process for employers and institutions. The deployment of 5G networks and its implications for mobile learning is another critical area for future research, with studies investigating how faster internet speeds and reduced latency can enhance the quality and interactivity of mobile learning experiences. Lastly, the impact of mobile learning on digital well-being should not be overlooked, with future research focusing on developing and evaluating tools that promote a healthy balance between technology use and overall well-being, including features that monitor and manage screen time and encourage regular breaks. While there is great potential for integrating smartphone and tablet applications into higher education, there are a number of issues that need to be resolved. By focusing on equitable access, personalized learning, immersive experiences, microlearning, gamification, blockchain, 5G, and digital well-being, future research can offer insightful information to improve the efficiency of mobile learning technology, which will eventually result in the creation of a more welcoming, interesting, and encouraging learning environment that gets students ready for success in the digital era.

# REFERENCES


Abed, R., & Alsalih, M. (2023). Learners' perception of e-learning during the COVID-19 pandemic: Electronic or remote education? Confrontation and challenges.

Agasi, K. (2024). Improving critical thinking by using technology in the learning process. *Interdisciplinary Journal of Research and Development, 11*(1 S1), 76. DOI: 10.56345/ijrdv11n1s112

Aggarwal, D., Sharma, D., & Saxena, A. (2024). Smart education: An emerging teaching pedagogy for interactive and adaptive learning methods. *Journal of Learning and Educational Policy, 4*(44), 1–9. DOI: 10.55529/jlep.44.1.9

Ariarta, A., Mu'in, M., Latif, M., Saadah, A., & Harini, S. (2024). Learning quality, time management and social support on student academic achievement. *EDUKASIA: Jurnal Pendidikan dan Pembelajaran, 5*, 11–22. DOI: 10.62775/edukasia.v5i2.778

Bradley, V. M. (2021). Learning management system (LMS) use with online instruction. *International Journal of Technology in Education.* https://doi.org/DOI: 10.46328/ijte.36

Caudill, J. (2007). The growth of m-learning and the growth of mobile computing: Parallel developments. *International Review of Research in Open and Distance Learning, 8*(2). Advance online publication. DOI: 10.19173/irrodl.v8i2.348

Chilaka, O. (2024). Embracing the power of technology: A gateway to innovation and connectivity.

Cojean, S., & Grand, M. (2024). Note-taking by university students on paper or a computer: Strategies during initial note-taking and revision. *The British Journal of Educational Psychology, 94*(2), 557–570. Advance online publication. DOI: 10.1111/bjep.12663 PMID: 38302257

Croucher, G. (2019). *Productivity and higher education.* Oxford Bibliographies., DOI: 10.1093/obo/9780199756810-0227

Dahri, N., Yahaya, N., Al-Rahmi, W., Almogren, A., & Vighio, M. (2024). Investigating factors affecting teachers' training through mobile learning: Task–technology fit perspective. *Education and Information Technologies*. Advance online publication. DOI: 10.1007/s10639-023-12434-9

Dembe, A., & Extension, K. P. (2024). The integration of virtual reality (VR) and augmented reality (AR) in classroom settings, *3*, 102–113.

Dhawan, S. (2020). Online learning: A panacea in the time of COVID-19 crisis. *Journal of Educational Technology Systems*, *49*(1), 5–22. Advance online publication. DOI: 10.1177/0047239520934018

Furqon, M., Sinaga, P., Liliasari, L., & Riza, L. (2023). The impact of learning management system (LMS) usage on students. *TEM Journal, 12*, 1082–1089. https://doi.org/DOI: 10.18421/TEM122-54

Ghoulam, K., Bouikhalene, B., Babori, A., & Noureddine, F. (2024). Exploring the impact of mobile devices in electronics e-learning: A case study evaluating the effectiveness of mobile learning applications in the field of electronics and sensors. *Advances in Mobile Learning Educational Research*, *4*(2), 1058–1072. DOI: 10.25082/AMLER.2024.02.001

Halder, I., Halder, S., & Guha, A. (2015). Educational use of mobile phones by undergraduate students: An Indian perspective. *Online Journal of Communication and Media Technologies*, *5*(4). Advance online publication. DOI: 10.29333/ojcmt/2526

Hidayati, T., & Diana, S. (2019). Students' motivation to learn English using mobile applications: The case of Duolingo and Hello English. [Journal of English Education and Linguistics Studies]. *JEELS*, *6*(2), 189–213. DOI: 10.30762/jeels.v6i2.1233

Hinze, A., Vanderschantz, N., Timpany, C., Cunningham, S., Saravani, S.-J., & Wilkinson, C. (2022). A study of mobile app use for teaching and research in higher education. *Technology. Knowledge and Learning*, *28*(3), 1271–1299. Advance online publication. DOI: 10.1007/s10758-022-09599-6

Horvers, A., Molenaar, I., West, H., Bosse, T., & Lazonder, A. (2024). Multimodal measurements enhance insights into emotional responses to immediate feedback. *Frontiers in Psychology, 14*, 1294386. Advance online publication. DOI: 10.3389/fpsyg.2023.1294386 PMID: 38362521

Husnita, L., Rahayuni, A., Fusfitasari, Y., Siswanto, E., & Rintaningrum, R. (2023). The role of mobile technology in improving accessibility and quality of learning. *Al-Fikrah: Jurnal Manajemen Pendidikan.*

Indra, M., Sutarto, S., Kharizmi, M., Nurmiati, A., & Susanto, A. (2023). Optimizing the potential of technology-based learning increases student engagement. *Al-Fikrah: Jurnal Manajemen Pendidikan, 11*(2), 233. DOI: 10.31958/jaf.v11i2.10554

Inyutina, A. I., & Kartashova, V. N. (2023). Modern LMS systems as a component of the organization of hybrid education in higher educational institutions. *Educational Psychology in Polycultural Space, 61*(1), 81–89. DOI: 10.24888/2073-8439-2023-61-1-81-89

Jian, M. (2023). Personalized learning through AI. *Advances in Engineering Innovation, 5*(1), 16–19. Advance online publication. DOI: 10.54254/2977-3903/5/2023039

Jose, D. (2024). Data privacy and security concerns in AI-integrated educational platforms. *Recent Trends in Management and Commerce, 5*(2), 87–91. DOI: 10.46632/rmc/5/2/19

Joshi, M., Shen, Z., & Kausar, S. (2023). *Enhancing inclusive education by leveraging artificial intelligence technologies for personalized support and accessibility in special education for students with diverse learning needs.*

Kessler, M., Loewen, S., & Gonulal, T. (2023). Mobile-assisted language learning with Babbel and Duolingo: Comparing L2 learning gains and user experience. *Computer Assisted Language Learning*. Advance online publication. DOI: 10.1080/09588221.2023.2215294

Lavidas, K., Athanassopoulos, S., Priovolou, K., Lappa, A., & Voutsina, L. (2020). Investigating practices and perceptions of higher education students concerning the utilization of mobile devices in their studies. *Journal of Open and Distance Education, 15*(2), 114. Advance online publication. DOI: 10.12681/jode.21990

Liu, X. (2022). The effects of tablet use on student learning achievements, participation, and motivation at different levels. *International Journal of Technology-Enhanced Education, 1*(1), 1–17. DOI: 10.4018/IJTEE.304819

Machado, A., Rodrigues dos Santos, J., Sacavém, A., & Sousa, M. (2024). Digital transformations: Artificial intelligence in higher education.

Mealings, K., Miles, K., & Buchholz, J. (2024). The classroom communication and collaboration appraisal: A comprehensive tool for assessing communication and collaboration in classroom environments. *Journal of Speech, Language, and Hearing Research: JSLHR, 67*(6), 1–20. DOI: 10.1044/2024_JSLHR-23-00586 PMID: 38683060

Meccawy, M. (2022). Creating an immersive XR learning experience: A roadmap for educators. *Electronics (Basel), 11*(21), 3547. Advance online publication. DOI: 10.3390/electronics11213547

Mesuwini, J., & Mokoena, S. (2024). Exploring online teaching and learning challenges for the technical and vocational education and training lecturer. *Journal of Education and e-learning Research, 11*(1), 193–202. DOI: 10.20448/jeelr.v11i1.5423

Modlinger, D. (2024). Mobile learning. In *Mobile learning* (Chapter 6). DOI: 10.1007/978-3-658-44695-6_6

Muhammad Nazeef, N., Khan, A., & Ali, J. (2024). Impact of collaborative learning on students' academic performance in teacher education programs. *Journal of Asian Development Studies, 13*(1), 1054–1068. DOI: 10.62345/jads.2024.13.1.87

Nasir, A., Nasrum, A., Haeniah, N., & Haeniah, N. (2023). Utilization of Chamilo LMS in learning media courses as an effort to digitalize education. *International Journal of Business, Law, and Education, 4*(2), 1394–1407. DOI: 10.56442/ijble.v4i2.284

Oluwaseyi, J., & Cena, J. (2024). Effectiveness of e-learning platforms, learning management systems, and mobile apps in modern education.

Open University. (2024). *The Open University*. https://www.open.ac.uk

Papadopoulos, P., Zygouris, N., Koziri, M., Loukopoulos, T., & Stamoulis, G. (2018). Mobivoke: A mobile system architecture to support off-school collaborative learning processes. DOI: 10.1007/978-3-319-75175-7_57

Penn State University. (2023). *The Pennsylvania State University*. https://www.psu.edu

Polat, D., & Gödek, Y. (2021). PhET interactive simulations: Virtual laboratory use in science teaching.

Qasim, S. (2024). Beyond the classroom: Emerging technologies to enhance learning.

Quadri, N. N., Choudhary, H., Ahmad, N., & Alqahtani, J. (2023). Mobile learning in higher education: A systematic literature review.

Rizvi, N. (2024). Unleashing the power of AI avatars: Revolutionizing education for the digital generation.

Salhab, R., & Daher, W. (2023). University students' engagement in mobile learning. *European Journal of Investigation in Health, Psychology and Education*, *13*(1), 202–216. DOI: 10.3390/ejihpe13010016 PMID: 36661766

Silvestre, A., & Jacinto, H. (2021). The use of technologies in mathematics teaching and learning in Portugal: Teachers' pre-pandemic views. DOI: 10.21125/edulearn.2021.1848

Sougleridi, E., Kopsidas, S., Vavougios, D., Avramopoulos, A., & Kanapitsas, A. (2024). Embedding AI into LMS and eLearning platforms. DOI: 10.1007/978-3-031-51979-6_38

Stanford University. (2022). *Center for Teaching and Learning*. https://ctl.stanford.edu

Swailes, N. (2016). A virtual laboratory and game-based app to facilitate self-directed online learning in histology. *The FASEB Journal*, *30*(S1). Advance online publication. DOI: 10.1096/fasebj.30.1_supplement.568.15

Tejada-Simon, M. (2024). Exploring the impact of game-based learning and creative active-learning activities on student engagement and academic performance: A case study in the basic sciences for pharmacy education. *Physiology (Bethesda, MD)*, *39*(S1), 935. Advance online publication. DOI: 10.1152/physiol.2024.39.S1.935

Teodorescu, C. A., Durnoi, A. N. C., & Vargas, V. M. (2023). The rise of the mobile Internet: Tracing the evolution of portable devices. In *Proceedings of the International Conference on Business Excellence* (Vol. 17, No. 1, pp. 1645–1654). Sciendo. DOI: 10.2478/picbe-2023-0147

University of Central Florida. (2018). *University of Central Florida*. https://www.ucf.edu

Yang, Z., & Zhong, S. (2023). Task offloading and resource allocation for edge-enabled mobile learning. *China Communications*, *20*(4), 326–339. DOI: 10.23919/JCC.fa.2022-0521.202304

Ymchenko, A. (2022). Changing the paradigm of blended learning in the system of classical education in conditions of the active use of LMS Moodle and Google Classroom. *Scientific Journal of Khortytsia National Academy*, 112–129. DOI: 10.51706/2707-3076-2022-7-12

# APPENDIX

*Table 2. Abbreviations*

| **AI** | **Artificial Intelligence** |
|---|---|
| AR | Augmented Reality |
| ASU | Arizona State University |
| LMS | Learning Management Systems |
| ML | Mobile Learning |
| PC | Personal computer |
| PDA | Personal Digital Assistant |
| SMS | Short Message Service |
| UCF | University of Central Florida |
| UK | United Kingdom |
| VR | Virtual reality |
| 5G | 5th Generation |

# Compilation of References


Abdulameer Nayyef Aldabbagh, Y., & Ibrahim Zghair Alnasrawi, A. (2025). A Comparative Analysis of Classification Algorithms in Land Cover Mapping: A Study of MLC, Mahalanobis Distance, NNC, and SVM for 2023 and 2018: Case study of Wasit province, central Iraq. *Jordanian Journal of Informatics and Computing*, *2025*(1), 49–66. DOI: 10.63180/jjic.thestap.2025.1.6

Abed, R., & Alsalih, M. (2023). Learners' perception of e-learning during the COVID-19 pandemic: Electronic or remote education? Confrontation and challenges.

About, S. W. A. Y. A. M. (n.d.). *Swayam Central*. https://swayam.gov.in/about

Abu Laila, D., Aljawarneh, M., Al-Na'amneh, Q., & Bin Sulaiman, R. (2025). Optimizing Intrusion Detection Systems through Benchmarking of Ensemble Classifiers on Diverse Network Attacks. *STAP Journal of Security Risk Management*, *2025*(1), 71–84. DOI: 10.63180/jsrm.thestap.2025.1.4

Addula, S. R., Norozpour, S., & Amin, M. (2025). Risk Assessment for Identifying Threats, vulnerabilities and countermeasures in Cloud Computing. *Jordanian Journal of Informatics and Computing*, *2025*(1), 38–48. DOI: 10.63180/jjic.thestap.2025.1.5

Adipat, S., Laksana, K., Busayanon, K., Ausawasowan, A., & Adipat, B. (2021). Engaging students in the learning process with game-based learning: The fundamental concepts. *International Journal of Technology in Education*, *4*(3), 542–552. DOI: 10.46328/ijte.169

Agasi, K. (2024). Improving critical thinking by using technology in the learning process. *Interdisciplinary Journal of Research and Development*, *11*(1 S1), 76. DOI: 10.56345/ijrdv11n1s112

Aggarwal, D., Sharma, D., & Saxena, A. (2024). Smart education: An emerging teaching pedagogy for interactive and adaptive learning methods. *Journal of Learning and Educational Policy*, *4*(44), 1–9. DOI: 10.55529/jlep.44.1.9

Aithal, P. S., & Aithal, S. (2016). Impact of online education on higher education system. *International Journal of Engineering Research and Modern Education*, *1*(1), 225–235.

Akyüz, H. I., & Samsa, S. (2009). The effects of a blended learning environment on the critical thinking skills of students. *Procedia: Social and Behavioral Sciences*, *1*(1), 1744–1748. Advance online publication. DOI: 10.1016/j.sbspro.2009.01.308

Albinhamad, H., Alotibi, A., Alagnam, A., Almaiah, M., & Salloum, S. (2025). Vehicular Ad-hoc Networks (VANETs): A Key Enabler for Smart Transportation Systems and Challenges. *Jordanian Journal of Informatics and Computing*, *2025*(1), 4–15. DOI: 10.63180/jjic.thestap.2025.1.2

Al-Emran, M. (2023). Mobile apps support blended learning adoption in Middle Eastern higher education. *Education and Information Technologies*, *28*(4), 4215–4232. DOI: 10.1007/s10639-022-11431-4

Alfares, N. S. (2025). Investigating the efficacy of WordWall platform in enhancing vocabulary learning in Saudi EFL classroom. *International Journal of Game-Based Learning*, *15*(1), 1–12. DOI: 10.4018/IJGBL.367870

Alghareeb, M. S., & Almaayah, M. (2025). Cyber Security Risk Management for Threats in Wireless LAN: A Literature Review. *STAP Journal of Security Risk Management*, *2025*(1), 22–58. DOI: 10.63180/jsrm.thestap.2025.1.2

Alharthi, M. A. (2021). The effectiveness of virtual classrooms as an alternative to traditional classrooms during the COVID-19 pandemic: Problems and solutions. *Life Science Journal*, *18*(5), 24–32. https://www.researchgate.net/publication/370679821_The_Effectiveness_of_Virtual_Classrooms_as_an_Alternative_to_Traditional_Classrooms_during_the_Covid-19_pandemic_Problems_and_Solutions

Al-Hunaiyyan, A., Alhajri, R., & Bimba, A. T. (2020). Smart learning environments: Concepts and implementations. *Education and Information Technologies*, *25*(4), 2783–2805.

Ali, A., Khan, R. M. I., & Alouraini, A. (2023). A Comparative Study on the Impact of Online and Blended Learning. *SAGE Open*, *13*(1), 1–10. DOI: 10.1177/21582440231154417

Alismaiel, O. A. (2022). Develop a New Model to Measure the Blended Learning Environments Through Students' Cognitive Presence and Critical Thinking Skills. *International Journal of Emerging Technologies in Learning*, *17*(12), 150–169. Advance online publication. DOI: 10.3991/ijet.v17i12.30141

Al-Kahtani, M. A., Aljuaid, R., & Alshmrany, A. (2021). *A Systematic Review of Mobile Learning Applications for Online Learning*. NAER Journal.

Ally, M., & Samaka, M. (2018). *E-learning, M-learning, and D-learning: Conceptual definition and comparative analysis*. SAGE Journals.

Almaayah, M., & Sulaiman, R. B. (2024). Cyber Risk Management in the Internet of Things: Frameworks, Models, and Best Practices. *STAP Journal of Security Risk Management*, *2024*(1), 3–23. DOI: 10.63180/jsrm.thestap.2024.1.1

Almahasees, Z., Mohsen, K., & Amin, M. O. (2021). Faculty's and Students' Perceptions of Online Learning During COVID-19. *Frontiers in Education*, *6*, 638470. https://slejournal.springeropen.com/articles/10.1186/s40561-023-00225-1. DOI: 10.3389/feduc.2021.638470

Almaiah, M. A., Al-Khasawneh, A., & Althunibat, A. (2020). Exploring the critical challenges and factors influencing the E-learning system usage during COVID-19 pandemic. *Education and Information Technologies*, *25*(6), 5261–5280. DOI: 10.1007/s10639-020-10219-y PMID: 32837229

Almaiah, M. A., Ayouni, S., Hajjej, F., Lutfi, A., Almomani, O., & Awad, A. B. (2022). Smart Mobile Learning Success Model for Higher Educational Institutions in the Context of the COVID-19 Pandemic. *Electronics (Basel)*, *11*(8), 1–13. DOI: 10.3390/electronics11081278

Almaiah, M. A., Jalil, M. A., & Man, M. (2016a). Extending the TAM to examine the effects of quality features on mobile learning acceptance. *Journal of Computers in Education*, *3*(4), 453–485. DOI: 10.1007/s40692-016-0074-1

Almaiah, M. A., Jalil, M. A., & Man, M. (2016b). Preliminary study for exploring the major problems and activities of mobile learning system: A case study of Jordan. *Journal of Theoretical and Applied Information Technology*, *93*(2), 580–594.

Almarshood, R., & Rahman, M. M. H. (2025). Enhancing Intrusion Detection Systems by Using Machine Learning in Smart Cities: Issues, Challenges and Future Research Direction. *STAP Journal of Security Risk Management*, *2025*(1), 3–21. DOI: 10.63180/jsrm.thestap.2025.1.1

Almelehy, B., Anwar, A., Nassreddine, G., & Maayah, M. (2025). Web Application Security: An Analytical Study of Cyber Threats and Defense Mechanisms. *Journal of Cyber Security and Risk Auditing*, *2025*(3), 43–62. DOI: 10.63180/jcsra.thestap.2025.3.5

Al-Na'amneh, Q., Aljawarneh, M., Alhazaimeh, A. S., Hazaymih, R., & Shah, S. M. (2025). Securing Trust: Rule-Based Defense Against On/Off and Collusion Attacks in Cloud Environments. *STAP Journal of Security Risk Management*, *2025*(1), 85–114. DOI: 10.63180/jsrm.thestap.2025.1.5

Alotaibi, A., Aldawghan, H., & Rahman, M. M. H. (2026). IoT Security Concerns with Non-Fungible Tokens: A Review. *STAP Journal of Security Risk Management*, *2026*(1), 1–30. DOI: 10.63180/jsrm.thestap.2026.1.1

Al-Shareeda, M. A., & Hawi, I. A. (2026). IoT-Driven Enterprise Development and Supply Chain Risk Management in Healthcare Organizations. *STAP Journal of Security Risk Management, 2026*(1), 98–111. DOI: 10.63180/jsrm.thestap.2026.1.7

Al-Shareeda, M. A., Obaid, A. A., & Almajid, A. A. H. (2025). The Role of Artificial Intelligence in Bodybuilding: A Systematic Review of Applications, Challenges, and Future Prospects. *Jordanian Journal of Informatics and Computing, 2025*(1), 16–26. DOI: 10.63180/jjic.thestap.2025.1.3

Alshinwan, M., Memon, A. G., Ghanem, M. C., & Almaayah, M. (2025). Unsupervised text feature selection approach based on improved Prairie dog algorithm for the text clustering. *Jordanian Journal of Informatics and Computing, 2025*(1), 27–36. DOI: 10.63180/jjic.thestap.2025.1.4

America, N. (2023). *Artificial Intelligence in Schools: Privacy and Security Considerations.* Retrieved from https://newamerica.org/oti/blog/artificial-intelligence-in-schools-privacy-and-security-considerations/

Amponsah, S., Ussher, Y., & Benjamin, K. A. (2021). Availability and access to support services in a blended learning environment. *International Journal of Information and Communication Technology Education, 17*(1), 57–71. Advance online publication. DOI: 10.4018/IJICTE.2021010104

Anderson, T. (2010). *The theory and practice of online learning* (2nd ed.). Athabasca University Press.

Andronico, P., Minutoli, S., & Kuruoglu, E. E. (2014). Understanding elderly needs for designing a digitally extended environment via tablets. In Paternò, F., Santoro, C., & Tarantino, L. (Eds.), *Human-computer interaction* (pp. 284–287). Springer., DOI: 10.1007/978-3-319-07854-0_33

Ansari, M. S., & Tripathi, A. (2021). An investigation of effectiveness of mobile learning apps in higher education in India. *ResearchGate.* https://www.researchgate.net/publication/349731965_An_Investigation_of_Effectiveness_of_Mobile_Learning_Apps_in_Higher_Education_in_India

Apoko, T. W. (2023). EFL Preservice Teachers' Intrinsic Motivation in Writing Essays in a Blended Learning Environment. *VELES: Voices of English Language Education Society*, *7*(1), 126–134. DOI: 10.29408/veles.v7i1.7838

Aremu, B. V., & Adeoluwa, O. V. (2021). M-learning: A nexus for adult learners' motivation and readiness to learn in federal universities at South-west, Nigeria. *Journal of Digital Educational Technology*, *2*(1), ep2201. Advance online publication. DOI: 10.21601/jdet/11361

Ariarta, A., Mu'in, M., Latif, M., Saadah, A., & Harini, S. (2024). Learning quality, time management and social support on student academic achievement. *EDUKASIA: Jurnal Pendidikan dan Pembelajaran, 5*, 11–22. DOI: 10.62775/edukasia.v5i2.778

Arnold, K. E., & Pistilli, M. D. (2012). Course signals at Purdue: Using learning analytics to increase student success. *Proceedings of the 2nd International Conference on Learning Analytics and Knowledge*, 267–270.

Audah Kareem, K., & Al-Shareeda, M. A. (2026b). A Cost-Effective Embedded Weather Monitoring System for Airport Air Traffic Control Applications. *Jordanian Journal of Informatics and Computing*, *2026*(1), 38–46. DOI: 10.63180/jjic.thestap.2026.1.4

Awedh, M., Mueen, A., Zafar, B., & Manzoor, U. (2015). Using Socrative and smartphones for the support of collaborative learning. *International Journal of Integrating Technology in Education*, *3*(4), 15–25. DOI: 10.5121/ijite.2014.3402

Bacca, J., Baldiris, S., Fabregat, R., & Graf, S., & Kinshuk. (2014). Augmented reality trends in education. *Computers & Education*, *72*, 66–81.

Baker, R. S., & Inventado, P. S. (2014). Educational data mining and learning analytics. In *Learning analytics* (pp. 61–75). Springer. DOI: 10.1007/978-1-4614-3305-7_4

Baker, R., & Smith, L. (2019). *Educ-AI-tion rebooted? Exploring the future of artificial intelligence in schools and colleges*. Nesta.

Balfour, S. P. (2013). Assessing writing in MOOCs: Automated essay scoring and calibrated peer review. *Research & Practice in Assessment, 8*(1), 40–48.

Bates, T. (2015). *Teaching in a digital age: Guidelines for designing teaching and learning*. Tony Bates Associates Ltd.

Behera, S. K. (2013). E- and M-Learning: A Comparative Study. *International Journal on New Trends in Education and Their Implications*.

Bergström, P. (2018). Investigating University Students' Intention to Use Mobile Learning Management Systems in Sweden. *Innovations in Education and Teaching International, 55*(5), 576–586. https://www.tandfonline.com/doi/full/10.1080/14703297.2018.1557068

Berkovich, I. (2021). The impact of AI-powered tutors on student engagement. *Educational Technology Research and Development, 69*(2), 89–104.

Bienkowski, M., Feng, M., & Means, B. (2012). *Enhancing teaching and learning through educational data mining and learning analytics*. U.S. Department of Education.

Billinghurst, M., & Duenser, A. (2012). Augmented reality in the classroom. *Computer, 45*(7), 56–63. DOI: 10.1109/MC.2012.111

Binns, R. (2018). Fairness in Machine Learning: Lessons for AI in Education. *AI & Society, 33*(1), 95–108.

Binns, R. (2018). Fairness in machine learning: Lessons from political philosophy. *Proceedings of the 2018 Conference on Fairness, Accountability, and Transparency, 81*, 149–159.

Black, P., & Wiliam, D. (1998). Assessment and classroom learning. *Assessment in Education: Principles, Policy & Practice, 5*(1), 7–74. DOI: 10.1080/0969595980050102

Bloom, B. S., Engelhart, M. D., Furst, E. J., Hill, W. H., & Krathwohl, D. R. (1956). *Taxonomy of educational objectives: The classification of educational goals*. Longmans.

Bradley, V. M. (2021). Learning management system (LMS) use with online instruction. *International Journal of Technology in Education.* https://doi.org/DOI: 10.46328/ijte.36

Braun, V., & Clarke, V. (2021). One size fits all? What counts as quality practice in (reflexive) thematic analysis? *Qualitative Research in Psychology, 18*(3), 328–352. DOI: 10.1080/14780887.2020.1769238

Brown, T. H. (2005). Towards a model for m-learning in Africa. *International Journal on E-Learning, 4*(3), 299–315.

Brusilovsky, P., & Millán, E. (2007). User models for adaptive hypermedia and adaptive educational systems. *The Adaptive Web, 4321*, 3–53. DOI: 10.1007/978-3-540-72079-9_1

Burstein, J., Chodorow, M., & Leacock, C. (2003). Criterion: Online essay evaluation: An application for automated evaluation of student essays. *AI Magazine, 24*(3), 27–36.

Campbell University. (2025, March 6). *AI in higher education: A summary of recent surveys of students and faculty.* Academic Technology Blog. https://sites.campbell.edu/academictechnology/2025/03/06/ai-in-higher-education-a-summary-of-recent-surveys-of-students-and-faculty/

Caudill, J. (2007). The growth of m-learning and the growth of mobile computing: Parallel developments. *International Review of Research in Open and Distance Learning, 8*(2). Advance online publication. DOI: 10.19173/irrodl.v8i2.348

Caudill, J. (2007). *The growth of m-learning and the growth of mobile computing: Parallel developments.* ResearchGate.

CEDTECH. (2024). *An Examination of Mobile Learning Adoption in Higher Education Research Trends in Twenty Years.* Retrieved from https://www.cedtech.net/article/an-examination-of-mobile-learning-adoption-in-higher-education-research-trends-in-twenty-years-15612

Chandran, A. L., Samual, J., Safavi, S., & Ali, A. (2025). A Comparative Analysis of Machine Learning Models on Detecting Malware in Android Devices. *Journal of Cyber Security and Risk Auditing*, *2025*(4), 327–346. DOI: 10.63180/jcsra.thestap.2025.4.10

Cheng, Q., Li, B., & Zhou, Y. (2021). Research on evaluation system of classroom teaching quality in colleges and universities based on 5G environment. In *Proceedings of the 2021 International Conference on Control and Intelligent Robotics* (pp. 74–85). ACM. DOI: 10.1145/3473714.3473728

Chen, X., Xie, H., Hwang, G. J., & Wang, H. (2020). A systematic review of artificial intelligence-driven educational technologies. *Journal of Educational Technology & Society*, *23*(2), 1–15.

Chervonyi, P., Alieksieieva, S., Senchylo-Tatlilioglu, N., Chastnyk, O., & Diomidova, N. (2021). Peculiarities of the higher education development in a blended learning environment. *Revista on Line de Política e Gestão Educacional*, *25*(3), 2241–2257. DOI: 10.22633/rpge.v25i3.15959

Chilaka, O. (2024). Embracing the power of technology: A gateway to innovation and connectivity.

Cojean, S., & Grand, M. (2024). Note-taking by university students on paper or a computer: Strategies during initial note-taking and revision. *The British Journal of Educational Psychology*, *94*(2), 557–570. Advance online publication. DOI: 10.1111/bjep.12663 PMID: 38302257

Creswell, J. W., & Plano Clark, V. L. (2018). *Designing and conducting mixed methods research* (3rd ed.). SAGE.

Crompton, H. (2013). A historical overview of mobile learning: Toward learner-centered education. In Berge, Z. L., & Muilenburg, L. Y. (Eds.), *Handbook of mobile learning* (pp. 3–14). Routledge.

Croucher, G. (2019). *Productivity and higher education.* Oxford Bibliographies., DOI: 10.1093/obo/9780199756810-0227

Cuong, L. H., Thi, N., Giang, H., Thi, N., & Tu, T. (2022). Design of an Educational Game for Blended Learning Environment. *JST: Smart Systems and Devices*, *32*(3), 17–25. DOI: 10.51316/jst.160.ssad.2022.32.3.3

Dahri, N., Yahaya, N., Al-Rahmi, W., Almogren, A., & Vighio, M. (2024). Investigating factors affecting teachers' training through mobile learning: Task–technology fit perspective. *Education and Information Technologies*. Advance online publication. DOI: 10.1007/s10639-023-12434-9

Davarasan, A., Samual, J., Palansundram, K., & Ali, A. (2024). A Comprehensive Review of Machine Learning Approaches for Android Malware Detection. *Journal of Cyber Security and Risk Auditing*, *2024*(1), 39–60. DOI: 10.63180/jcsra.thestap.2024.1.5

Davis, F. D. (1989). Perceived usefulness, perceived ease of use, and user acceptance of information technology. *Management Information Systems Quarterly*, *13*(3), 319–340. DOI: 10.2307/249008

Dembe, A., & Extension, K. P. (2024). The integration of virtual reality (VR) and augmented reality (AR) in classroom settings, *3*, 102–113.

Dhawan, S. (2020). Online learning: A panacea in the time of COVID-19 crisis. *Journal of Educational Technology Systems*, *49*(1), 5–22. Advance online publication. DOI: 10.1177/0047239520934018

Dichev, C., & Dicheva, D. (2017). Gamifying education: What is known and what remains uncertain. *International Journal of Educational Technology in Higher Education*, *14*(9), 1–21.

Digital Education Council. (2025). What faculty want: Key results from the Global AI Faculty Survey 2025. *Digital Education Council*. https://www.digitaleducationcouncil.com/post/what-faculty-want-key-results-from-the-global-ai-faculty-survey-2025

Digital Media Use and Mental Health. (2023).

Economides, A. A., & Nikolaou, N. (2008). Evaluation of handheld devices for mobile learning. *International Journal of Engineering Education*, *24*(1), 3–13.

Educational Data Mining. (2024). *Developing Explainable AI Systems to Support Feedback for Students.* Retrieved from https://educationaldatamining.org/edm2024/proceedings/2024.EDM-doctoral-consortium.125/index.html

Edu, T., Negricea, C., Zaharia, R., & Zaharia, R. M. (2022). Factors influencing student transition to online education in the COVID 19 pandemic lockdown: Evidence from Romania. *Ekonomska Istrazivanja, 35*(1), 3291–3304. DOI: 10.1080/1331677X.2021.1990782

ERIC. (2023). *Collaborative mobile learning and its impact on student communication and motivation.* Retrieved from https://files.eric.ed.gov/fulltext/EJ1441219.pdf

ERIC. (2024). *Exploratory Analysis of Smart Partnerships in Education: Risks and Opportunities.* Retrieved from https://eric.ed.gov/?id=EJ1107145

ERIC. (2024). *The Effect of Mobile Learning Applications on Students' Academic Achievement and Attitudes.* Retrieved from https://files.eric.ed.gov/fulltext/EJ1174817.pdf

Eve, M. P. (2014). *Open access and the humanities.* Cambridge University Press. DOI: 10.1017/CBO9781316161012

Ferdig, R. (2020). Virtual reality and augmented reality in education. *Journal of Online Learning Research, 6*(1), 45–60.

Fernández-Batanero, J. M., Montenegro-Rueda, M., Fernández-Cerero, J., & Tadeu, P. (2022). Online education in higher education: emerging solutions in crisis times. In *Heliyon* (Vol. 8, Issue 8). DOI: 10.1016/j.heliyon.2022.e10139

Floridi, L., & Cowls, J. (2019). A unified framework of five principles for AI in society. *Harvard Data Science Review, 1*(1). Advance online publication. DOI: 10.1162/99608f92.8cd550d1

Furqon, M., Sinaga, P., Liliasari, L., & Riza, L. (2023). The impact of learning management system (LMS) usage on students. *TEM Journal, 12*, 1082–1089. https://doi.org/DOI: 10.18421/TEM122-54

Garrison, D. R., Anderson, T., & Archer, W. (2021). The community of inquiry framework: Ten years later. *Internet and Higher Education, 13*(1), 3–9.

Ghoulam, K., Bouikhalene, B., Babori, A., & Noureddine, F. (2024). Exploring the impact of mobile devices in electronics e-learning: A case study evaluating the effectiveness of mobile learning applications in the field of electronics and sensors. *Advances in Mobile Learning Educational Research, 4*(2), 1058–1072. DOI: 10.25082/AMLER.2024.02.001

Gikas, J., & Grant, M. M. (2013). Mobile computing devices in higher education: Student perspectives. *The Internet and Higher Education, 19*, 18–26. DOI: 10.1016/j.iheduc.2013.06.002

Gligorea, I., Cioca, M., Oancea, R., Gorski, A.-T., Gorski, H., & Tudorache, P. (2023). Adaptive learning using artificial intelligence in e-learning: A literature review. *Education Sciences, 13*(12), 1216. DOI: 10.3390/educsci13121216

Gobrecht, A., Tuma, F., Möller, M., Zöller, T., Zakhvatkin, M., Wuttig, A., Sommerfeldt, H., & Schütt, S. (2024). Beyond human subjectivity and error: a novel AI grading system. *arXiv preprint arXiv:2405.04323*. Retrieved from https://arxiv.org/abs/2405.04323

Goggin, G., & Newell, C. (2020). *Digital disability: The social construction of disability in new media*. Routledge.

Government of India. (2020). *National Education Policy 2020*. Ministry of Education. https://www.education.gov.in/nep2020

Graesser, A. C. (2016). Conversations with AutoTutor help students learn. *International Journal of Artificial Intelligence in Education, 26*(1), 124–132. DOI: 10.1007/s40593-015-0086-4

Graesser, A. C., Conley, M. W., & Olney, A. (2012). Intelligent tutoring systems. In Harris, K. R., Graham, S., & Urdan, T. (Eds.), APA educational psychology handbook: Vol. 3. *Applications to learning and teaching* (pp. 451–473). American Psychological Association.

Grant-Kels, J. M. (2017). The ethical arguments in support of open access journals. *International Journal of Women's Dermatology*, *3*(1), 4–5. DOI: 10.1016/j.ijwd.2017.01.002 PMID: 28492047

Grant, W. C. (1993). Wireless Coyote: A computer-supported field trip. *Communications of the ACM*, *36*(5), 57–59. DOI: 10.1145/155049.155062

Grech, A., & Camilleri, A. F. (2017). *Blockchain in education*. Publications Office of the European Union.

Halder, I., Halder, S., & Guha, A. (2015). Educational use of mobile phones by undergraduate students: An Indian perspective. *Online Journal of Communication and Media Technologies*, *5*(4). Advance online publication. DOI: 10.29333/ojcmt/2526

Han, H., & Trimi, S. (2022). Cloud computing-based higher education platforms during the COVID-19 pandemic. *arXiv preprint arXiv:2203.03714*. DOI: 10.1145/3514262.3514307

Hanif, H., Hamzah, M., & Harun, R. N. S. R. (2022). Optimizing Participation with Helpers to Enhance Interaction in a Blended Learning Environment. *International Journal of Emerging Technologies in Learning*, *17*(17), 155–170. Advance online publication. DOI: 10.3991/ijet.v17i17.26415

Hao, S., Dennen, V. P., & Mei, L. (2017). Influential factors for mobile learning acceptance among Chinese users. *Educational Technology Research and Development*, *65*(1), 101–123. DOI: 10.1007/s11423-016-9465-2

Hardan, A., & Ghabayen, M. (2025). Artificial Intelligence in Auditing: Transforming Processes for Enhanced Effectiveness. *STAP International Journal of Accounting and Business Intelligence*, *2025*(1), 24–38.

Hartt, M., Hosseini, H., & Mostafapour, M. (2020). Game on: Exploring the effectiveness of game-based learning. *Planning Practice and Research*, *35*(5), 589–604. DOI: 10.1080/02697459.2020.1778859

Hidayat, A., Iqbal, M., & Faizal, M. (2021). Predicting Students' Use of Mobile-Learning Management Systems: An Extension of UTAUT2. *The Journal of Educators Online*, *18*(1), n1. https://files.eric.ed.gov/fulltext/EJ1383890.pdf

Hidayati, T., & Diana, S. (2019). Students' motivation to learn English using mobile applications: The case of Duolingo and Hello English. [Journal of English Education and Linguistics Studies]. *JEELS*, *6*(2), 189–213. DOI: 10.30762/jeels.v6i2.1233

Himmelstein, D. S., Romero, A. R., Levernier, J. G., Munro, T. A., McLaughlin, S. R., Tzovaras, B. G., & Greene, C. S. (2018). Sci-Hub provides access to nearly all scholarly literature. *eLife*, *7*, e32822. DOI: 10.7554/eLife.32822 PMID: 29424689

Hinze, A., Vanderschantz, N., Timpany, C., Cunningham, S., Saravani, S.-J., & Wilkinson, C. (2022). A study of mobile app use for teaching and research in higher education. *Technology. Knowledge and Learning*, *28*(3), 1271–1299. Advance online publication. DOI: 10.1007/s10758-022-09599-6

Hirschberg, J., & Manning, C. D. (2015). Advances in natural language processing. *Science*, *349*(6245), 261–266. DOI: 10.1126/science.aaa8685 PMID: 26185244

Holmes, W., Bialik, M., & Fadel, C. (2019). Artificial intelligence in education: Promises and implications for teaching and learning. *Center for Curriculum Redesign*. https://oro.open.ac.uk/60255/

Holmes, W., Bialik, M., & Fadel, C. (2019). *Artificial intelligence in education: Promises and implications for teaching and learning*. Center for Curriculum Redesign.

Ho, M., Ang, S., Huy, S., & Janarthanan, M. (2026). Cybersecurity Risks and Challenges in Smart Cities: A Review with Insights for Cambodia. *STAP Journal of Security Risk Management*, *2026*(1), 87–97. DOI: 10.63180/jsrm.thestap.2026.1.6

Ho, M., Ang, S., Huy, S., & Janarthanan, M. (2026). MUMSPI: A Model for Usability Measurement of Single-Platform Interface for Multi-Tasking in Big Data Tools. *Jordanian Journal of Informatics and Computing*, *2026*(1), 1–14. DOI: 10.63180/jjic.thestap.2026.1.1

Hort, M., & Burri, T. (2019). Fair and unbiased algorithmic decision-making: Current state and future challenges. *ResearchGate*. https://www.researchgate.net/publication/330409917_Fair_and_Unbiased_Algorithmic_Decision_Making_Current_State_and_Future_Challenges

Horvers, A., Molenaar, I., West, H., Bosse, T., & Lazonder, A. (2024). Multimodal measurements enhance insights into emotional responses to immediate feedback. *Frontiers in Psychology*, *14*, 1294386. Advance online publication. DOI: 10.3389/fpsyg.2023.1294386 PMID: 38362521

Huang, C. H., Tsai, Y. S., & Pardo, A. (2020). Artificial intelligence in education: An overview of the current state of AI applications in higher education. *Computers & Education*, *151*, 103872.

Husnita, L., Rahayuni, A., Fusfitasari, Y., Siswanto, E., & Rintaningrum, R. (2023). The role of mobile technology in improving accessibility and quality of learning. *Al-Fikrah: Jurnal Manajemen Pendidikan*.

Hussein, M., & Nordin, N. (2023). Leveraging AI in mobile learning to support education: A taxonomy of AI applications. *Journal of Innovations in Digital Pedagogy*, *3*(1), 45–62. DOI: 10.12345/jidp.v3i1.7347

Huy, S., Ang, S., Ho, M., & Balasubramaniam, V. (2026). Securing API Ecosystems in Banking: A Critical Review of Cyber Risks, Control Frameworks, and Future Trends. *Jordanian Journal of Informatics and Computing*, *2026*(1), 25–37. DOI: 10.63180/jjic.thestap.2026.1.3

Hwang, G.-J., Xie, H., Wah, B. W., & Gasevic, D. (2020). Vision, challenges, roles, and research issues of artificial intelligence in education. *Computers & Education: Artificial Intelligence*, *1*, 100001.

Ibrahim, M. (2024). AI as a collaborative partner: Fostering peer feedback and cooperation for higher education literacy. In *AI approaches to literacy in higher education* (pp. 117-149). https://doi.org/DOI: 10.4018/979-8-3693-1054-0.ch006

Ibrahim, U. (2024). Assessing the impact of mobile applications on student engagement in ICT and computer science education. *International Journal of Applied and Advanced Multidisciplinary Research*, *2*(3), 221–232. https://www.researchgate.net/publication/379525078_Assessing_the_Impact_of_Mobile_Applications_on_Student_Engagement_in_ICT_and_Computer_Science_Education. DOI: 10.59890/ijaamr.v2i3.1533

IEEE. (2024). Personalized Adaptive Learning Technologies Based on Machine Learning Techniques. *IEEE Xplore*. Retrieved from https://ieeexplore.ieee.org/document/10124747

IEEE. (2024). *Real-Time Learning Analytics Dashboard for Students in Online Classes*. Retrieved from https://ieeexplore.ieee.org/document/9368340

Indra, M., Sutarto, S., Kharizmi, M., Nurmiati, A., & Susanto, A. (2023). Optimizing the potential of technology-based learning increases student engagement. *Al-Fikrah: Jurnal Manajemen Pendidikan*, *11*(2), 233. DOI: 10.31958/jaf.v11i2.10554

Inyutina, A. I., & Kartashova, V. N. (2023). Modern LMS systems as a component of the organization of hybrid education in higher educational institutions. *Educational Psychology in Polycultural Space*, *61*(1), 81–89. DOI: 10.24888/2073-8439-2023-61-1-81-89

Irwanto, I., Saputro, A. D., Widiyanti, W., & Laksana, S. D. (2022). *Global Trends on Mobile Learning in Higher Education: A Bibliometric Analysis (2002–2022)*. ResearchGate. Retrieved from https://www.researchgate.net/publication/368308070_Global_Trends_on_Mobile_Learning_in_Higher_Education_A_Bibliometric_Analysis_2002-2022

Jaafar, H. S., Abed, A. A., & Al-Shareeda, M. A. (2026). A Secure Industrial Internet of Things (IIoT) Framework for Real-Time PI Control and Cloud-Integrated Industrial Monitoring. *STAP Journal of Security Risk Management, 2026*(1), 77–86. DOI: 10.63180/jsrm.thestap.2026.1.5

Jia, L. (2022). Clustering Analysis of the Evaluation Data on Students' Learning under a Blended Environment. *International Journal of Emerging Technologies in Learning*, *17*(23), 128–144. Advance online publication. DOI: 10.3991/ijet.v17i23.35941

Jian, M. (2023). Personalized learning through AI. *Advances in Engineering Innovation*, *5*(1), 16–19. Advance online publication. DOI: 10.54254/2977-3903/5/2023039

Joan, D. R. R. (2013). *A Study on Mobile Learning as a Learning Style in Modern Education.* i-Manager's Journal on School Educational Technology, 8(4), 31-38. Retrieved from https://files.eric.ed.gov/fulltext/EJ1098313.pdf

Johnson, M., Smith, R., & Brown, L. (2021). *Emerging technologies in education: The role of artificial intelligence*. Academic Press.

Jonassen, D. H. (1999). *Designing constructivist learning environments.* Routledge.

Jones, M., & Miller, R. (2021). The effects of AI tutors on critical thinking skills. *Educational Psychology Journal*, *46*(2), 150–168.

Jose, D. (2024). Data privacy and security concerns in AI-integrated educational platforms. *Recent Trends in Management and Commerce*, *5*(2), 87–91. DOI: 10.46632/rmc/5/2/19

Joshi, M., Shen, Z., & Kausar, S. (2023). *Enhancing inclusive education by leveraging artificial intelligence technologies for personalized support and accessibility in special education for students with diverse learning needs.*

Journal, N. A. E. R. (2024). *Systematic Review on the Development of Mobile Learning Applications for Online Education.* Retrieved from https://naerjournal.com/article/view/v12n2-1

Kaliisa, R., & Picard, M. (2024). The promise of mobile learning for higher education in Sub-Saharan Africa. *International Journal of Educational Technology in Higher Education*, *21*(2), 45–62. DOI: 10.1186/s41239-024-00456-7

Kamalov, F., Santandreu Calonge, D., & Gurrib, I. (2023). New era of artificial intelligence in education: Towards a sustainable multifaceted revolution. *Sustainability (Basel)*, *15*(16), 12451. DOI: 10.3390/su151612451

Kareem, K., & Al-Shareeda, M. A. (2026). A Low-Complexity Li-Fi Communication Framework for Short-Range Text Transmission. *Jordanian Journal of Informatics and Computing*, *2026*(1), 15–24. DOI: 10.63180/jjic.thestap.2026.1.2

Keegan, D. (2002). The future of learning: From e-learning to m-learning. *International Review of Research in Open and Distance Learning.*

Kessler, M., Loewen, S., & Gonulal, T. (2023). Mobile-assisted language learning with Babbel and Duolingo: Comparing L2 learning gains and user experience. *Computer Assisted Language Learning*. Advance online publication. DOI: 10.1080/09588221.2023.2215294

Koedinger, K. R., Corbett, A. T., & Perfetti, C. (2013). The knowledge-learning-instruction framework: Bridging the science-practice chasm to enhance robust student learning. *Cognitive Science*, *37*(4), 757–798. PMID: 22486653

Kukulska-Hulme, A., & Traxler, J. (2019). *Mobile learning: A handbook for educators and trainers*. Routledge.

Kukulska-Hulme, A., & Traxler, J. (2019). *Mobile learning: The next generation*. Routledge.

Kumar, A., & Sharma, S. (2020). 5G and mobile learning: The next phase of digital education. *Journal of Educational Technology & Society*, *23*(2), 45–57.

Kumar, A., Singh, J., & Gupta, R. (2020). Artificial intelligence in education: The role of intelligent tutoring systems. *Computers & Education*, *149*, 103842.

Kumar, V., & Sharma, R. (2021). AI-driven learning systems: Adoption and impact on higher education. *Journal of Educational Computing Research*, *59*(5), 823–845.

Lakulu, M. M., Shihan, A., Muttalib, M., & Zainuddin, N. (2024). Leveraging AI in mobile learning to support education: A taxonomy of AI applications. *Journal of Infrastructure. Policy and Development*, *8*(16), 7347. DOI: 10.24294/jipd7347

Laurillard, D. (2013). *Rethinking university teaching: A conversational framework for the effective use of learning technologies* (2nd ed.). Routledge. DOI: 10.4324/9781315012940

Lavidas, K., Athanassopoulos, S., Priovolou, K., Lappa, A., & Voutsina, L. (2020). Investigating practices and perceptions of higher education students concerning the utilization of mobile devices in their studies. *Journal of Open and Distance Education*, *15*(2), 114. Advance online publication. DOI: 10.12681/jode.21990

Liang, P., Bommasani, R., Lee, T., Tsipras, D., Soylu, D., Yasunaga, M., Zhang, Y., Narayanan, D., Wu, Y., Kumar, A., Newman, B., Yuan, B., Yan, B., Zhang, C., Cosgrove, C., Manning, C. D., Ré, C., Acosta-Navas, D., Hudson, D. A., & Koreeda, Y. (2022). Holistic evaluation of language models. *arXiv preprint arXiv:2211.09110*. https://doi.org//arXiv.2211 .09110DOI: 10.48550

Liang, P., Lee, M., & Chang, H. (2022). Algorithmic bias in AI-based educational technologies. *Computers & Education*, *170*, 104–115.

Li, F., & Wang, C. (2023). Artificial intelligence and edge computing for teaching quality evaluation based on 5G-enabled wireless communication technology. *Journal of Cloud Computing (Heidelberg, Germany)*, *12*(45), 45. Advance online publication. DOI: 10.1186/s13677-023-00418-6

Liu, W., & Yu, H. (2012). Effectiveness study of English learning in blended learning environment. *Theory and Practice in Language Studies*, *2*(3), 524–530. DOI: 10.4304/tpls.2.3.524-530

Liu, X. (2022). The effects of tablet use on student learning achievements, participation, and motivation at different levels. *International Journal of Technology-Enhanced Education*, *1*(1), 1–17. DOI: 10.4018/IJTEE.304819

Long, N. T., & Van Hanh, N. (2020). A structural equation model of blended learning culture in the classroom. *International Journal of Higher Education*, *9*(4), 99. Advance online publication. DOI: 10.5430/ijhe.v9n4p99

Luckin, R., Holmes, W., Griffiths, M., & Forcier, L. B. (2016). *Intelligence unleashed: An argument for AI in education*. Pearson Education.

Luo, H., Feng, A., & Wang, Y. (2019). AI-powered grading systems: A case study on automated assessment of written responses. *Computers & Education*, *142*, 103635.

Machado, A., Rodrigues dos Santos, J., Sacavém, A., & Sousa, M. (2024). Digital transformations: Artificial intelligence in higher education.

Machumu, H., Zhu, C., & Almasi, M. (2018). Students' motivational factors and engagement strategies in constructivist- based blended learning environments. *Afrika Focus*, *31*(1), 13–34. DOI: 10.1163/2031356X-03101002

Manuel. (2023, December 15). 8 game based learning examples. Teach Find. https://teachfind.com/teaching-strategies/game-based-learning/

Marienko, M., Nosenko, Y., & Shyshkina, M. (2020). Personalization of learning using adaptive technologies and augmented reality. *arXiv preprint arXiv:2011.05802*. DOI: 10.31812/123456789/4418

MDPI. (2023). AI in Collaborative Learning and Automated Assessments. *MDPI.com*. Retrieved from https://www.mdpi.com/2071-1050/15/16/12451

MDPI. (2024). *Evaluating Recent Advances in Affective Intelligent Tutoring Systems*. Retrieved from https://www.mdpi.com/2227-7102/14/8/839

Mealings, K., Miles, K., & Buchholz, J. (2024). The classroom communication and collaboration appraisal: A comprehensive tool for assessing communication and collaboration in classroom environments. *Journal of Speech, Language, and Hearing Research: JSLHR*, *67*(6), 1–20. DOI: 10.1044/2024_JSLHR-23-00586 PMID: 38683060

Means, B., Toyama, Y., Murphy, R., Bakia, M., & Jones, K. (2013). The Effectiveness of Online and Blended Learning: A Meta-Analysis of the Empirical Literature. *Teachers College Record*, *115*(3), 1–47. DOI: 10.1177/016146811311500307

Meccawy, M. (2022). Creating an immersive XR learning experience: A roadmap for educators. *Electronics (Basel)*, *11*(21), 3547. Advance online publication. DOI: 10.3390/electronics11213547

Mehta, S., & Shah, H. (2020). Digital divide and online education in India. *Journal of Educational Planning and Administration*, *34*(2), 123–138.

Mese, C., & Dursun, O. O. (2019). Effectiveness of gamification elements in blended learning environments. *Turkish Online Journal of Distance Education*, *20*(3), 119–142. DOI: 10.17718/tojde.601914

Mesuwini, J., & Mokoena, S. (2024). Exploring online teaching and learning challenges for the technical and vocational education and training lecturer. *Journal of Education and e-learning Research*, *11*(1), 193–202. DOI: 10.20448/jeelr.v11i1.5423

Mhammed, K. (2025). Cyber Intelligence & Moroccan National Security: What Strategy for Managing and Mitigating Cyber Threats against Future Sporting Events. *Journal of Cyber Security and Risk Auditing*, *2025*(4), 366–383. DOI: 10.63180/jcsra.thestap.2025.4.12

Ministry of Education. (2020). *DIKSHA: Digital Infrastructure for Knowledge Sharing*. Government of India.

Ministry of Education. (2023). *All India Survey on Higher Education (AISHE) 2022–23*. Government of India.

Mittmann, G., Barnard, A., Zehetmayer, S., Wimmer, S., Zehetner, V., Doerfler, S., Woodcock, K., & Schrank, B. (2024). A serious game for emotion regulation in adolescents. *International Journal of Game-Based Learning, 14*(1), 1–17. DOI: 10.4018/IJGBL.358940

Modlinger, D. (2024). Mobile learning. In *Mobile learning* (Chapter 6). DOI: 10.1007/978-3-658-44695-6_6

Molnar, G., & Szuts, Z. (2022). Use of artificial intelligence in electronic learning environments. In G. Molnár (Ed.), *IEEE 5th International Conference and Workshop in Óbuda on Electrical and Power Engineering (CANDO-EPE 2022)* (pp. 137-140). https://doi.org/DOI: 10.1109/CANDO-EPE57516.2022.10046356

Monde, L. (2024). *MIA Seconde: The AI Tutor for National Education in France.* Retrieved from https://www.lemonde.fr/pixels/article/2024/10/29/mia-seconde-a-l-heure-de-chatgpt-le-tuteur-sous-intelligence-artificielle-de-l-education-nationale-est-il-deja-obsolete_6364449_4408996.html

Montclair (2024). *Adaptive Learning Strategies and Practices.* Retrieved from https://www.montclair.edu/itds/digital-pedagogy/pedagogical-strategies-and-practices/adaptive-learning/

Moodle (2023). *Top 5 Learning Management Systems in 2023.* Retrieved from https://moodle.com/news/top-5-learning-management-systems-in-2023/

Mousa, R. S., & Shehab, R. (2025). Applying risk analysis for determining threats and countermeasures in workstation domain. *Journal of Cyber Security and Risk Auditing, 2025*(1), 12–21. DOI: 10.63180/jcsra.thestap.2025.1.2

Muhammad Nazeef, N., Khan, A., & Ali, J. (2024). Impact of collaborative learning on students' academic performance in teacher education programs. *Journal of Asian Development Studies, 13*(1), 1054–1068. DOI: 10.62345/jads.2024.13.1.87

Muthuprasad, T., Aiswarya, S., Aditya, K. S., & Jha, G. K. (2021). Students' perception and preference for online education in India during COVID-19 pandemic. *Social Sciences & Humanities Open*, *3*(1), 100101. DOI: 10.1016/j.ssaho.2020.100101 PMID: 34173507

Nasir, A., Nasrum, A., Haeniah, N., & Haeniah, N. (2023). Utilization of Chamilo LMS in learning media courses as an effort to digitalize education. *International Journal of Business, Law, and Education*, *4*(2), 1394–1407. DOI: 10.56442/ijble.v4i2.284

Nearpod (2023). *Interactive Learning with Nearpod.* Retrieved from https://nearpod.com/

Nguyen, T., Lee, J., & Chen, Y. (2025). Equity gaps in mobile learning adoption: Evidence from an Asian multi-site study. *Computers & Education*, *221*, 105122. DOI: 10.1016/j.compedu.2025.105122

O'Shea, M. (2023). *The Evolution of Technology in the Classroom.* Purdue University.

Oluwaseyi, J., & Cena, J. (2024). Effectiveness of e-learning platforms, learning management systems, and mobile apps in modern education.

Open University. (2024). *The Open University*. https://www.open.ac.uk

Pachler, N., Bachmair, B., & Cook, J. (2010). *Mobile learning: Structures, agency, practices*. Springer. DOI: 10.1007/978-1-4419-0585-7

Panda, S., & Mishra, S. (2018). Mobile Learning: Challenges for Teachers of Indian Open Universities. *Journal of Learning for Development*, *5*(1), 35–46. https://jl4d.org/index.php/ejl4d/article/download/145/147/745

Pan, L., Tlili, A., Li, J., Jiang, F., Shi, G., Yu, H., & Yang, J. (2021). How to implement game-based learning in a smart classroom? A model based on a systematic literature review and Delphi method. *Frontiers in Psychology*, *12*, 749837. Advance online publication. DOI: 10.3389/fpsyg.2021.749837 PMID: 34925153

Papadopoulos, P., Zygouris, N., Koziri, M., Loukopoulos, T., & Stamoulis, G. (2018). Mobivoke: A mobile system architecture to support off-school collaborative learning processes. DOI: 10.1007/978-3-319-75175-7_57

Pardo, A., Mirriahi, N., & Dawson, S. (2016). Identifying student learning strategies in MOOCs through learning analytics. *Journal of Learning Analytics*, *3*(3), 99–115.

Patterson, J. M. (2024). Repurposing video games as discussion tools. *International Journal of Game-Based Learning*, *14*(1), 1–19. DOI: 10.4018/IJGBL.349589

PCMag. (2023). *The Best LMS (Learning Management Systems)*. Retrieved from https://www.pcmag.com/picks/the-best-lms-learning-management-systems

Pedraja-Rejas, L., Muñoz-Fritis, C., Rodríguez-Ponce, E., & Laroze, D. (2024). Mobile learning and its effect on learning outcomes and critical thinking: A systematic review. *Applied Sciences (Basel, Switzerland)*, *14*(19), 9105. DOI: 10.3390/app14199105

Pedro, F., Subosa, M., Rivas, A., & Valverde, P. (2019). Artificial intelligence in education: Challenges and opportunities for sustainable development. *UNESCO*. https://unesdoc.unesco.org/ark:/48223/pf0000366994

Pedró, F., Subosa, M., Rivas, A., & Valverde, P. (2019). *Working papers on education policy (No. 7) [Programme and meeting document]*. UNESCO. ED-2019/WS/8.

Penn State University. (2023). *The Pennsylvania State University*. https://www.psu.edu

Pew Research. (2021). *Some Digital Divides Persist Between Rural, Urban, and Suburban America*. Retrieved from https://www.pewresearch.org/short-reads/2021/08/19/some-digital-divides-persist-between-rural-urban-and-suburban-america/

Piaget, J. (1950). *The psychology of intelligence*. Routledge.

PLOS ONE (2024). *Adoption of Mobile Learning in University Contexts: A Systematic Review.* Retrieved from https://journals.plos.org/plosone/article?id=10.1371%2Fjournal.pone.0304116

PMC. (2023). *The Role of Mobile-Computer-Supported Collaborative Learning in Enhancing Communication and Engagement.* Retrieved from https://pmc.ncbi.nlm.nih.gov/articles/PMC5613807/

Polat, D., & Gödek, Y. (2021). PhET interactive simulations: Virtual laboratory use in science teaching.

Price, D., & Winchester, M. (2025). Blended learning equity in higher education: A scoping review. *Australian Educational Researcher*. Advance online publication. DOI: 10.1007/s13384-025-00859-9

Privacy International. (2020). *Financial report 2019-2020.* Privacy International. https://privacyinternational.org/financial-report/4210/financial-report-2019-2020

Prohorets, E., & Plekhanova, M. (2015). Interaction intensity levels in blended learning environment. *Procedia: Social and Behavioral Sciences, 174*(3822), 3818–3823. DOI: 10.1016/j.sbspro.2015.01.1119

Puentedura, R. R. (2006). Transformation, technology, and education. *Hippasus Blog*. https://hippasus.com/resources/tte/

Puspitasari, D. (2021). BLENDED LEARNING ENVIRONMENT: PROMOTING LEARNERS' SELF-EFFICACY. *ELTALL: English Language Teaching. Applied Linguistic and Literature*, *2*(1), 39–46. DOI: 10.21154/eltall.v2i1.2729

Qahmash, A. (2018). The Potentials of Using Mobile Technology in Teaching Individuals with Learning Disabilities: A Review of Special Education Technology Literature. *TechTrends*, *62*(6), 647–653. Advance online publication. DOI: 10.1007/s11528-018-0298-1

Qasim, S. (2024). Beyond the classroom: Emerging technologies to enhance learning.

Quadri, N. N., Choudhary, H., Ahmad, N., & Alqahtani, J. (2023). Mobile learning in higher education: A systematic literature review.

Radianti, J., Majchrzak, T. A., Fromm, J., & Wohlgenannt, I. (2020). A systematic review of immersive virtual reality applications for higher education. *Computers & Education, 147*, 103778. DOI: 10.1016/j.compedu.2019.103778

Raftr (2025). *Mobile-first student engagement in higher education: From apps to AI-powered platforms*. https://www.raftr.com/mobile-first-student-engagement-in-higher-education/

Ramesh, H., Ismail, N., Abd Rahman, N. A., & Ali, A. (2026). PhishGuard: AI-Driven Graph-Based Analysis for Smarter Email Security. *STAP Journal of Security Risk Management, 2026*(1), 31–45. DOI: 10.63180/jsrm.thestap.2026.1.2

Ramesh, S., & Dutta, P. (2022). AI and the Challenge of Linguistic Inclusion in Education. *Computational Linguistics Journal, 44*(3), 78–99.

Reich, J. (2020). *Failure to disrupt: Why technology alone can't transform education*. Harvard University Press.

ResearchGate. (2024). *Adaptive Learning: A Cluster-Based Literature Review (2011-2022)*. Retrieved from https://www.researchgate.net/publication/374084207_Adaptive_learning_a_cluster-based_literature_review_2011-2022

ResearchGate. (2024). *Adoption of Mobile Technology for Mobile Learning by University Students During COVID-19*. Retrieved from https://www.researchgate.net/publication/353271720_Adoption_of_mobile_technology_for_mobile_learning_by_university_students_during_COVID-19

ResearchGate. (2024). An Analysis of the Research on Adaptive Learning. *ResearchGate.net*. Retrieved from https://www.researchgate.net/publication/234778584_An_analysis_of_the_research_on_adaptive_Learning_The_next_generation_of_e-learning

ResearchGate. (2024). Relationship Between Student Engagement and Academic Performance. *ResearchGate.net*. Retrieved from https://www.researchgate.net/publication/381430411_Relationship_between_student_engagement_and_academic_performance

ResearchGate. (2024). Technology Education. *Historical Research*.

Rienties, B., Cross, S., & Marsh, S. (2016). Harnessing learning analytics to identify student needs. *The Internet and Higher Education*, *29*, 1–10.

Rivera, J. H. (2017). The Blended Learning Environment: A Viable Alternative for Special Needs Students. *Journal of Education and Training Studies*, *5*(2), 79–84. DOI: 10.11114/jets.v5i2.2125

Rizvi, N. (2024). Unleashing the power of AI avatars: Revolutionizing education for the digital generation.

Roff, K. A. (2018). Student Satisfaction and/or Dissatisfaction in Blended Learning Environments. *Frontiers in Education Technology*, *1*(2), 149. DOI: 10.22158/fet.v1n2p149

Rogers, E. M. (2003). *Diffusion of innovations* (5th ed.). Free Press.

Ruokonen, I., & Ruismäki, H. (2016). E-Learning in Music: A Case Study of Learning Group Composing in a Blended Learning Environment. *Procedia: Social and Behavioral Sciences*, *217*, 109–115. DOI: 10.1016/j.sbspro.2016.02.039

Sajja, R., Reddy, P. K., & Kumar, A. (2023). Artificial intelligence-enabled intelligent assistant for personalized and adaptive learning in higher education. *arXiv preprint*. https://arxiv.org/abs/2309.10892

Salhab, R., & Daher, W. (2023). University students' engagement in mobile learning. *European Journal of Investigation in Health, Psychology and Education*, *13*(1), 202–216. DOI: 10.3390/ejihpe13010016 PMID: 36661766

Selwyn, N. (2020a). The risks of digital education: From policy to practice. *Learning, Media and Technology*, *45*(2), 115–129.

Selwyn, N. (2020b). *Should robots replace teachers? AI and the future of education*. Polity Press.

Selwyn, N. (2022). The future of AI and education: Some cautionary notes. *European Journal of Education*, *57*(4), 620–631. Advance online publication. DOI: 10.1111/ejed.12532

Seman, S. A. A., Hashim, M. J., Roslin, R. M., & Ishar, N. I. M. (2019). Millennial learners' acceptance and satisfaction of blended learning environment. *Asian Journal of University Education*, *15*(3), 129. Advance online publication. DOI: 10.24191/ajue.v15i3.7845

Sergeeva, O. V., Zheltukhina, M. R., Ishmuradova, I. I., Kondakchian, N. A., Erokhova, N. S., & Zhdanov, S. P. (2025). A meta-analysis of the effectiveness of mobile-supported collaborative learning. *Online Journal of Communication and Media Technologies*, *15*(1), e202508. DOI: 10.30935/ojcmt/15948

Settles, B., & Meeder, B. (2016). A trainable spaced repetition model for language learning. *Proceedings of the 54th Annual Meeting of the Association for Computational Linguistics (*Volume 1*: Long Papers)*, 1848–1858. DOI: 10.18653/v1/P16-1174

Sharma, P. (2019). Digital revolution of Education 4.0. *International Journal of Engineering and Advanced Technology*, *9*(2), 3558–3564. DOI: 10.35940/ijeat.A1293.129219

Sharma, R., & Kumar, P. (2021). Adaptive e-learning environment based on learning styles and its impact on learners' performance. *International Journal of Educational Technology in Higher Education*, *18*(1), 1–24.

Sharma, R., & Kumar, P. (2024). *Smart Learning Environments: A Comprehensive Review on Artificial Intelligence and Learning Theories*. RSIS International Journal.

Sharma, S., Singh, G., Sharma, C. S., & Kapoor, S. (2024). Artificial intelligence in Indian higher education institutions: A quantitative study on adoption and perceptions. *International Journal of System Assurance Engineering and Management*, *15*(1), 123–135. DOI: 10.1007/s13198-023-02193-8

Shermis, M. D., & Burstein, J. C. (Eds.). (2013). *Handbook of automated essay evaluation: Current applications and new directions*. Routledge. DOI: 10.4324/9780203122761

Shishakly, R., Almaiah, M. A., Al-Otaibi, S., Lutfi, A., Alrawad, M., & Almulhem, A. (2023). A New Technological Model on Investigating the Utilization of Mobile Learning Applications: Extending the TAM. *Multimodal Technologies and Interaction*, *7*(9), 1–18. DOI: 10.3390/mti7090092

Shuja, A., Qureshi, I. A., Schaeffer, D. M., & Zareen, M. (2019). Effect of m-learning on students' academic performance mediated by facilitation discourse and flexibility. *Knowledge Management & E-Learning*, *11*(2), 158–200. DOI: 10.34105/j.kmel.2019.11.009

Shute, V. J., & Rahimi, S. (2021). Review of AI-driven formative assessment in education. *Computers & Education*, *166*, 104120.

Siemens, G. (2005). Connectivism: A learning theory for the digital age. *International Journal of Instructional Technology and Distance Learning*, *2*(1), 3–10.

Siemens, G. (2013). Learning analytics: The emergence of a discipline. *The American Behavioral Scientist*, *57*(10), 1380–1400. DOI: 10.1177/0002764213498851

Siemens, G., & Long, P. (2011). Learning analytics: The future of education. *EDUCAUSE Review*, *46*(1), 30–40.

Siemens, G., & Long, P. (2011). Penetrating the fog: Analytics in learning and education. *EDUCAUSE Review*, *46*(5), 30–32.

Silvestre, A., & Jacinto, H. (2021). The use of technologies in mathematics teaching and learning in Portugal: Teachers' pre-pandemic views. DOI: 10.21125/edulearn.2021.1848

Singh, N. (2019). Perception of the ODL Learners Towards the Use of Mobile Apps in Learning in India. *Asian Journal of Distance Education, 14*(2), 71–85. https://asianjde.com/ojs/index.php/AsianJDE/article/view/425

Smiderle, R., Rigo, S. J., Marques, L. B., Coelho, J. E. P., & Jaques, P. A. (2020). The impact of gamification on students' learning, engagement, and behavior based on their personality traits. *Smart Learning Environments, 7*(3), 3. Advance online publication. DOI: 10.1186/s40561-019-0098-x

Smith, J., & Brown, L. (2022). Generative artificial intelligence and web accessibility: Towards an inclusive and sustainable future. *ResearchGate.* https://www.researchgate.net/publication/383406745_Generative_Artificial_Intelligence_and_Web_Accessibility_Towards_an_Inclusive_and_Sustainable_Future

Smith, J., Lee, P., & Kumar, R. (2024). Barriers to AI adoption in education: A faculty perspective. *Journal of Science and Educational Research, 11*(8), 211–214. https://jsaer.com/download/vol-11-iss-8-2024/JSAER2024-11-8-211-214.pdf

Solutions, P. (2024). *Challenges in Mobile Learning Adoption.* Retrieved from https://www.paradisosolutions.com/blog/top-mobile-learning-benefits/

Song, H., Liu, Y., Li, K., Ding, L., Sun, X., & Zhang, X. (2020). Reform and future prospects in education based on 5G technology. In Amornpitoksuk, P. (Ed.), *Proceedings of the 3rd International Conference on Education Technology and Information System (ETIS 2020)* (pp. 136–142). Clausius Scientific Press.

Sougleridi, E., Kopsidas, S., Vavougios, D., Avramopoulos, A., & Kanapitsas, A. (2024). Embedding AI into LMS and eLearning platforms. DOI: 10.1007/978-3-031-51979-6_38

Spector, J. M. (2019). Emerging educational technologies and research directions. *Educational Technology Research and Development, 67*(4), 789–799.

Springer (2024). *Examining the Applications of Intelligent Tutoring Systems in Real-World Educational Settings: A Meta-Analysis.* Retrieved from https://link.springer.com/article/10.1007/s10639-022-11555-x

SpringerOpen. (2024). *Feedback Mechanisms in Adaptive Learning.* Retrieved from https://slejournal.springeropen.com/articles/10.1186/s40561-024-00292-y

Stanford University. (2022). *Center for Teaching and Learning.* https://ctl.stanford.edu

Sung, Y. T., Yang, J. M., & Lee, H. Y. (2017). The effects of mobile-computer-supported collaborative learning: Meta-analysis and critical synthesis. *Review of Educational Research, 87*(4), 768–805. DOI: 10.3102/0034654317704307 PMID: 28989193

Suresh Babu, C. V., & Auroshaa, A. (2024).,Influence of Data Analysis Tools in Teaching and Learning Economics. In *Innovative Pedagogy in the Digital Era: Empowering Learning Through Digital Leadership and Technology Integration* (pp. xx–xx). IGI Global

Suresh Babu, C. V., & Adhithya, S. (2023). ChatGPT and Other Generative AI Tools in Education: Transformative Potential and Ethical Considerations. In Keengwe, J. (Ed.), *Creative AI Tools and Ethical Implications in Teaching and Learning* (pp. 135–152). IGI Global., DOI: 10.4018/979-8-3693-0205-7.ch007

Suresh Babu, C. V., & Padma, R. (2023). Technology Transformation Through Skilled Teachers in Teaching Accountancy. In González-Lezcano, R. (Ed.), *Advancing STEM Education and Innovation in a Time of Distance Learning* (pp. 211–233). IGI Global., DOI: 10.4018/978-1-6684-5053-6.ch011

Suresh Babu, C. V., & Rohan, B. (2023). Evaluation and Quality Assurance for Rapid E-Learning and Development of Digital Learning Resources. In Santally, M., Rajabalee, Y., & Rajputh, R. (Eds.), *Implementing Rapid E-Learning Through Interactive Materials Development* (pp. 139–170). IGI Global., DOI: 10.4018/978-1-6684-4940-0.ch008

Swailes, N. (2016). A virtual laboratory and game-based app to facilitate self-directed online learning in histology. *The FASEB Journal, 30*(S1). Advance online publication. DOI: 10.1096/fasebj.30.1_supplement.568.15

SWAYAM. (2023). *SWAYAM annual report 2022–23*. Ministry of Education, Government of India.

Sweller, J. (1988). Cognitive load during problem solving: Effects on learning. *Cognitive Science, 12*(2), 257–285. DOI: 10.1207/s15516709cog1202_4

Tahir, R., & Wang, A. I. (2022). Evaluating the effectiveness of game-based learning for teaching refugee children Arabic using the integrated LEAGUÊ-GQM approach. *Behaviour & Information Technology, 43*(1), 110–138. DOI: 10.1080/0144929X.2022.2156386

Technologies, S. M. A. R. T. (2024). SMART and the University of Melbourne partner on neurodivergent research. *SMART Technologies*. https://www.smarttech.com/about/news/smart-and-the-university-of-melbourne-partner-on-neurodivergent-research

Tejada-Simon, M. (2024). Exploring the impact of game-based learning and creative active-learning activities on student engagement and academic performance: A case study in the basic sciences for pharmacy education. *Physiology (Bethesda, MD), 39*(S1), 935. Advance online publication. DOI: 10.1152/physiol.2024.39.S1.935

Tenopir, C., Christian, L., & Volentine, R. (2015). Scholarly Reading by Researchers: Summary Results from Surveys in Six Countries. *Learned Publishing, 28*(2), 124–132.

Teodorescu, C. A., Durnoi, A. N. C., & Vargas, V. M. (2023). The rise of the mobile Internet: Tracing the evolution of portable devices. In *Proceedings of the International Conference on Business Excellence* (Vol. 17, No. 1, pp. 1645–1654). Sciendo. DOI: 10.2478/picbe-2023-0147

The Verge. (2024). The quickly disappearing web. *The Verge.* https://www.theverge.com/24321569/internet-decay-link-rot-web-archive-deleted-culture

Top Hat. (2019, September 16). Game based learning definition and meaning. https://tophat.com/glossary/g/game-based-learning/

TPRE. (2023). Challenges in implementing mobile learning in higher education. *Teaching and Pedagogy Research Exchange.* https://tpre.ecu.edu/index.php/tpre/article/view/756

Traxler, J. (2018). Learning with mobile devices: The future of education. *Interactive Learning Environments, 26*(1), 4–16.

Treviranus, J. (2014). The Global Public Inclusive Infrastructure (GPII). *ResearchGate.* https://www.researchgate.net/publication/262400407_The_Global_Public_Inclusive_Infrastructure_GPII

UGC. (2021). *Guidelines on blended learning and evaluation reforms.* University Grants Commission.

UNESCO. (2021). *Education: From disruption to recovery. United Nations Educational, Scientific and Cultural Organization.* https://en.unesco.org/covid19/educationresponse

UNESCO. (2021). *Reimagining our futures together: A new social contract for education.* UNESCO Publishing. https://unesdoc.unesco.org/ark:/48223/pf0000379707

UNESCO. (2022). *Artificial intelligence in education: Challenges and opportunities for sustainable development.* United Nations Educational, Scientific and Cultural Organization.

University of Central Florida. (2018). *University of Central Florida.* https://www.ucf.edu

Van Dijk, J. (2020). *The Digital Divide*. Polity Press.

VanLehn, K. (2011). The relative effectiveness of human tutoring, intelligent tutoring systems, and other tutoring systems. *Educational Psychologist, 46*(4), 197–221. DOI: 10.1080/00461520.2011.611369

Vasanth, S., & Sumathi, C. (2020). Learning management systems through Moodle and Google Classroom for education. *Advances in Research, 21*(1), 32–37. DOI: 10.9734/air/2020/v21i1030249

Venkatesh, V., & Davis, F. D. (2000). A theoretical extension of the technology acceptance model. *Management Science, 46*(2), 186–204. DOI: 10.1287/mnsc.46.2.186.11926

Venkatesh, V., Morris, M. G., Davis, G. B., & Davis, F. D. (2003). User acceptance of information technology: Toward a unified view. *Management Information Systems Quarterly, 27*(3), 425–478. DOI: 10.2307/30036540

Vesselinov, R., & Grego, J. (2012). Duolingo effectiveness study. *City University of New York (CUNY) Report.*

Vij, A., & Agarwal, P. (2024). Assessing the challenges and opportunities of artificial intelligence in Indian education. *International Journal for Global Academic & Scientific Research, 3*(1), 36–44. DOI: 10.55938/ijgasr.v3i1.71

Vygotsky, L. S. (1978). *Mind in society: The development of higher psychological processes*. Harvard University Press.

Vygotsky, L. S. (1978). *Mind in society: The development of higher psychological processes*. Harvard University Press. Baker, R., & Smith, L. (2019). *Educ-AI-tion rebooted? Exploring the future of artificial intelligence in schools and colleges*. Nesta.

Vygotsky, L. S. (1978). *Mind in society: The development of higher psychological processes*. Harvard University Press.

Wang, A. I. (2015). The Wear Out Effect of a Game-Based Student Response System. *Computers & Education, 82*, 217–227. DOI: 10.1016/j.compedu.2014.11.004

Williamson, B., & Eynon, R. (2020). Algorithmic governance and the future of education: Artificial intelligence in learning analytics. *Learning, Media and Technology*, *45*(1), 23–36. DOI: 10.1080/17439884.2020.1798995

Williamson, B., & Eynon, R. (2020). Historical perspectives on learning analytics and AI in education. *Journal of Learning Analytics*, *7*(1), 40–58.

Wired. (2023, January 23). A major app flaw exposed the data of millions of Indian students. *Wired*. https://www.wired.com/story/india-diksha-app-data-leak

Wobbrock, J. O., Kane, S. K., Gajos, K. Z., Harada, S., & Froehlich, J. (2011). Ability-based design: Concept, principles, and examples. *ACM Transactions on Accessible Computing*, *3*(3), 9. Advance online publication. DOI: 10.1145/1952383.1952384

World Bank. (2023). *Digital transformation in higher education: Global trends and policy recommendations*. The World Bank Group.

Xiao, C., Ma, W., Song, Q., Xu, S. X., Zhang, K., Wang, Y., & Fu, Q. (2024). Human-AI Collaborative Essay Scoring: A Dual-Process Framework with LLMs. *arXiv preprint arXiv:2401.06431*. Retrieved from https://arxiv.org/abs/2401.06431

Xu, Z., Zhao, Y., Liew, J., Zhou, X., & Kogut, A. (2023). Synthesizing research evidence on self-regulated learning and academic achievement in online and blended learning environments: A scoping review. *Educational Research Review*, *39*, 1–17. DOI: 10.1016/j.edurev.2023.100510

Yang, C., & Taele, P. (2025). Audemy: AI for accessible education for blind students. *arXiv preprint*. https://arxiv.org/abs/2504.17117

Yang, Z., & Zhong, S. (2023). Task offloading and resource allocation for edge-enabled mobile learning. *China Communications*, *20*(4), 326–339. DOI: 10.23919/JCC.fa.2022-0521.202304

Yesmakhanova, Z., Zhussipova, G., Abjapparovna, K., Sumatokhin, S., Saparova, Z., & Assilbekova, B. (2022). Formation of digital competencies of future teachers of biology in a blended learning environment. *World Journal on Educational Technology: Current Issues*, *14*(5), 1612–1621. DOI: 10.18844/wjet.v14i5.8123

Ymchenko, A. (2022). Changing the paradigm of blended learning in the system of classical education in conditions of the active use of LMS Moodle and Google Classroom. *Scientific Journal of Khortytsia National Academy*, 112–129. DOI: 10.51706/2707-3076-2022-7-12

YouTube. (2023). *How to Use Kahoot! for Learning Engagement.* Retrieved from https://www.youtube.com/watch?v=07_sgCii_FE

Yuan, W. (2023). Perceptions of Chinese College Students towards Online Learning: Implications for Teaching in Blended Learning Environments. *The Educational Review, USA*, *7*(2), 185–194. Advance online publication. DOI: 10.26855/er.2023.02.009

Zawacki-Richter, O., Marín, V. I., Bond, M., & Gouverneur, F. (2019). Systematic review of research on artificial intelligence applications in higher education—Where are the educators? *International Journal of Educational Technology in Higher Education*, *16*(1), 39. Advance online publication. DOI: 10.1186/s41239-019-0171-0

Zhang, K., & Aslan, A. B. (2021). AI technologies for education: Recent research & future directions. *Computers and Education: Artificial Intelligence*, *2*, 100025. DOI: 10.1016/j.caeai.2021.100025

Zhang, L., & Aslan, B. (2021). Blockchain in higher education: A security perspective. *International Journal of Information Security*, *15*(2), 67–85.

Zhang, P., White, J., Schmidt, D., & Lenz, G. (2018). Blockchain-based credential verification in education. *Future Internet*, *10*(3), 27.

Zou, Y., Kuek, F., Feng, W., & Cheng, X. (2025). Digital learning in the 21st century: Trends, challenges, and innovations in technology integration. *Frontiers in Education*, *10*, 1562391. DOI: 10.3389/feduc.2025.1562391

# About the Contributors

**Mohammed Amin Almaiah** is an Associate Professor in the Department of Computer Science at University of Jordan. Almaiah is among the top 2% scientists in the world from 2020 up to now. He is working as Editor in Chief for the International Journal of Cybersecurity and Risk Assessment. He has published over 115 research papers in highly reputed journals such as the Engineering and Science Technology, an International Journal, Education and Information Technologies, IEEE Access and others. Most of his publications were indexed under the ISI Web of Science and Scopus. His current research interests include Cybersecurity, Cybersecurity-Risk Assessment and Blockchain .

**Said Salloum** hold a PhD in Computer Science, specializing in Artificial Intelligence and Cybersecurity, from the University of Salford, Manchester, UK. Notably, I have been recognized as one of the top 2% of scientists globally, a distinction affirmed in the Stanford University reports published in October 2021, 2022, and 2023.

***

**Vijay Ganpat Dhamore** is an accomplished academic and researcher with expertise in higher education, digital learning, and emerging pedagogical practices. His work focuses on integrating technology into education, with a strong emphasis on mobile learning, digital platforms, and ICT-enabled teaching methods tailored to the Indian higher education context. Dr. Dhamore has contributed significantly through research publications, case studies, and academic writings that explore the effectiveness of digital tools in enhancing accessibility, interactivity, and learner engagement. Dedicated to advancing knowledge, Dr. Dhamore actively engages in curriculum development, policy analysis, and scholarly writing to bridge the gap between traditional and modern teaching approaches. His academic contributions highlight the transformative potential of digital education in aligning with initiatives such as India's National Education Policy (NEP) 2020, thereby strengthening the role of technology in shaping future-ready education.

**D. Hepzibah Vinsyah Jeyaseeli** is currently working as Assistant Professor in the Department of Management Studies, Francis Xavier Engineering College, Tirunelveli .She has completed Ph.D. from Manonmaniam Sundaranar University and qualified UGC-NET in Management. She has published more than 10 research articles in various peer reviewed journals. She has published two books. She has 6 years of teaching experience.

**D.Joel Jebadurai** is working as a Associate professor in Department of MBA, St.Joseph's college of Engineering OMR Chennai Tamilnadu India. He has published more than 50 research articles in various peer reviewed journals. He has undertaken the funded projects and written two books also.

**Nikita Joshi Mishra** is working as an Associate Professor at IMS Ghaziabad, University courses campus. She has an experience of 14 years in academics and research. She was awarded her Ph.D. in Computer Science and Engineering from Amity University Uttar Pradesh. She completed her M.Tech (Computer Science & Engineering) from Amity University Uttar Pradesh and B.Tech from Uttar Pradesh Technical University. She has published 17 research papers in Scopus indexed journals and other reputed international journals. She has also been the reviewer of reputed IEEE international conferences and journals. Her research interest include image processing, data mining and machine learning, Computer Vision, Computer Graphics, Mobile Computing and Computer Networks She is a member of CSI, ACM, and Amity University Network Research lab.

**Divya Mathur** is an associate professor in Marwadi university. She has more than 30 years of experience. Her area of interest is AI, Marketing, AR And VR.

**M. Nanda Kumar** is an assistant professor at the department of management studies Gojan school of business and technology . with over two years of teaching experience and two years of industry expertise as an associate he bridges the gap between academia and industry His passion lies in Developing innovative management ideas fostering practical learning and nurturing future professionals into successful managers and leaders.

**Surenthran David P** is an Assistant Professor at Xavier Institute of Management & Entrepreneurship in Chennai, India. His academic contributions include publications in both Scopus and UGC CARE indexed journals, reflecting his active engagement in research and scholarly activities. In addition to his academic endeavors, he has also shared his expertise as a resource person in the field of Finance, contributing to the development and dissemination of knowledge in this crucial area.

**Tanya Shakya**, is currently pursuing Bachelor's of computer application from Institute of management studies, Ghaziabad (University Courses Campus). She has a keen interest in Data Science field and has done trainings and internships from different platforms.

**C.V. Suresh Babu** is a pioneer in content development. A true entrepreneur, he founded Anniyappa Publications, a company that is highly active in publishing books related to Computer Science and Management. Dr. C.V. Suresh Babu has also ventured into SB Institute, a center for knowledge transfer. He holds a Ph.D. in Engineering Education from the National Institute of Technical Teachers Training & Research in Chennai, along with seven master's degrees in various disciplines such as Engineering, Computer Applications, Management, Commerce, Economics, Psychology, Law, and Education. Additionally, he has UGC-NET/SET qualifications in the fields of Computer Science, Management, Commerce, and Education. Currently, Dr. C.V. Suresh Babu is a Professor in the Department of Information Technology at the School of Computing

Science, Hindustan Institute of Technology and Science (Hindustan University) in Padur, Chennai, Tamil Nadu, India.

**Monica Verna** is an Associate professor at Marwadi University. She is the head of the Commerce department and has 25 years of vast experience. Her area of interest is microfinance, banking, insurance and policy governance

**S. Vijayalakshmi** is a highly motivated and dedicated academic professional with over 4 years of experience as an Assistant Professor in Finance, Accounting, HR, Marketing and Logistics. she brings a unique blend of practical and theoretical knowledge to her teaching role. Currently working at Jaya College of Arts and Science. Her expertise includes academic leadership, student mentoring, and class counselling. She has successfully handled various responsibilities, including admission coordination, organizing industrial visits, Workshops, Seminars and Naan Mudhalvan courses. As an NPTEL coordinator, she has also facilitated online learning.

**Seema Yadav** has academic qualifications of M. Sc. (Chemistry), M Ed (RIE), UGC-NET, and PhD in Education. Presently she is working as an Assistant Professor, in the Department of Education at The Bhopal School of Education, Bhopal. She has almost 18 years of teaching experience. Her area of specialization is educational psychology, environmental education, digital learning, and curriculum development.

# Index

## A



## B



## D



## E



## G



## H